Electric System Operations

Evolving to the Modern Grid

Dr Vadari has a tremendous understanding of both the business and technical implications of Electric System Operations. Both business people and technical people need to understand how Electric System Operations will fit into the modern Electric Company's business. Dr Vadari's book provides that perspective.

Phil Crosby, ex-Accenture Partner, Global Utilities Industry

Few people have Mani's wide experience and the ability to take complexity into simplicity. His book explains all these elements, and should be mandatory reading for anyone interested in doing business in the electric industry.

Brian Dawson,
Executive VP of Business Development and Chairman of the Board,
Calico Energy Inc.

Peeling away each layer of complexity, Mani walks the reader through the evolution of electric operations. He then puts it all back together methodically to show how all the seemingly disparate pieces fit together. This is a book for the beginner, the practitioner and the expert.

Dr. Siri Varadan
Vice President Asset Management Solutions UISOL,
Formerly Sr. Principal Consultant,
Practice Area Coordinator (Asset Management in Utilities) KEMA

Mani Vadari's book, "Electric Systems Operations" illustrates in detail the importance of system operations within an electric utility and how it touches all aspects of the business as we advance in modernizing the grid. Mani maintains a practical approach throughout the book and drives home the importance of electric systems operations to the business of keeping the lights on. I recommend this book for everyone training or working in the utility business today as a vendor, partner, and employee or otherwise.

Michael Atkinson, Alstom Grid, North America, Vice-President

Until now, electric system operations have been more of a black box even within the utility and even more so externally. Dr. Mani Vadari has managed to navigate the reader through this complex part of the utility, explaining its evolution starting from the New York blackout, through deregulation to become the foundation of the modern grid. A serious book on the business of system operations is long overdue for our industry and Dr. Vadari has delivered to that mandate very well.

Guido Bartels,
Former GM, Energy & Utilities, IBM,
Chairman, Global Smart Grid Federation and
Immediate Past Chairman,
GridWise Alliance, Advisor, EnergyAid.org

In "Electric System Operations", Dr. Vadari provides a remarkably clear and extremely valuable guide to power system architecture and operations. This work covers all the key aspects of the modern electric grid including: basic definitions, the impact of deregulation, business insights, elucidation of the various IT-based management systems that continue to evolve, and concludes with his insight on future grid evolution. Dr. Vadari's work is mandatory reading for anyone with an interest in the modern electric grid.

James A. Misewich, Ph.D.
Associate Laboratory Director for Basic Energy Sciences
Brookhaven National Laboratory
Board of Directors, New York Battery and
Energy Storage Technology Consortium
Brookhaven Representative to Board of Directors,
New York State Smart Grid Consortium

Electric System Operations

Evolving to the Modern Grid

Mani Vadari

Library of Congress Cataloging-in-Publication Data
A catalog record for this book is available from the U.S. Library of Congress.

British Library Cataloguing in Publication Data
A catalogue record for this book is available from the British Library.

Cover design by Vicki Kane
Cover image courtesy of Midwest ISO.

ISBN 13: 978-1-60807-549-2

ARTECH HOUSE
685 Canton Street
Norwood, MA 02062

10 9 8 7 6 5 4 3 2 1

To Anu, Mayukha, and Akhi

I dedicate this book to my wife, Anu, who has patiently supported me in everything I have accomplished in my life, and without whom this book could not have become a reality. She has been my inspiration and she has even challenged my thinking in terms of the flow of information in this book.

Thanks a lot, Anu.

This book is also dedicated to my children, Mayukha and Akhi, who tolerated me during this entire period of writing this book and continued to shower their love and affection

Contents

Preface

I have always been fascinated and passionate about the subject of system operations. My first job was to develop a transient stability application for a dispatcher training simulator (DTS) scheduled to be delivered to New York Power Pool (NYPP) in 1989. However, when we delivered the DTS to NYPP, I found that there was much I needed to know about how DTSs are used to train system operators. Learning about system operators and how they do their job got me very interested with this important area of the electric utility and led to my interest in wanting to write this book.

The systems used by system operators are some of the most sophisticated in the utility and appear to be in a continuous state of transition and evolution. The system operator is responsible for keeping the grid up and running. Yet what happens inside the control center is not that well understood sometimes even within the utility.

This book is intended for people who work in a utility, or support the industry by providing solutions and services to them. It is also focused on the power system engineer who, for the most part will already have knowledge of the software systems and the algorithms used. It is designed as an introduction to this very important function and the business imperative that it delivers to. This book answers several important questions: (1) What is the business of system operations and why is it important? (2) What are the various systems that support this function and how do they integrate into the rest of the utility? (3) How does a utility's ability to operate its network contribute to sustaining its business in a viable and meaningful manner?

This book has been a labor of love for me for a very long time. After spending the early part of my career delivering energy management system (EMS) solutions to utilities, I found a dearth of books covering the entire breadth of EMS and associated systems. I started planning to write a book on EMSs from a technical aspect but going beyond just the algorithms—hardware, databases, mapboards, control centers, and so on.

Over time, my thoughts and approach evolved. I felt that I needed to expand the scope of the book to a broader focus on all aspects of system operations across the transmission, distribution, deregulation, smart grid, and the customer. My thoughts further evolved, moving me away from the purely technical to also include the business perspective. These modifications make this book somewhat unique in almost all respects. The book draws from my experiences and work over time delivering systems and solutions to utilities worldwide.

The book is organized in a way that brings the reader into the area of system operations by opening one door at a time. Starting with setting the stage with a definition of system operations, I follow it up with two chapters in which I analyze the impacts of deregulation and smart grid on system operations. Chapter 6 focuses on the business of system operations in which I present the people, process, technology, and strategy behind this area and how it contributes towards the utility's fundamental mandate of delivering reliable power. Chapter 7 defines the control center, which in many ways is the hallmark of system operations. The next four chapters focus on four key systems which are the foundational tools for the system operator: EMS, outage management system (OMS), distribution management system (DMS), and distributed energy management system (DEMS). DEMS is a new concept presented for the first time and covers the system operator and impact of the customer on this area. The book ends with a chapter on system operator training and the training simulator.

As a special note to students and practitioners in the field of system operations, this book discusses several powerful algorithms such as power flow, optimal power flow, three-phase unbalanced power flow, contingency analysis, and so forth. Several books are available in the marketplace that explain these algorithms in much more detail. I have specifically chosen to not go into detail in these areas because of the broader focus of what this book is trying to achieve.

Acknowledgments

Significant thanks are also due to Mrudhula Balasubramanyan, who was instrumental in doing some of the research that was necessary for the book and also for her due diligence with which she proofread every word and every sentence in every chapter to make sure they were all consistent in form and content. Her contribution has made this book that much better in both form and content and I owe a lot to her.

Thanks are also due to Gayle Wooster of Alstom Grid for responding patiently to all my requests, some of which were not that easy for her to get immediately, and even more so due to her busy schedule. My former colleagues at Accenture, especially David Rouls and Greg Smith, also deserve my thanks for their encouragement and support for the writing this book. It is all much appreciated.

No book of this magnitude can be developed without the support of several people who have helped me over the years and I will be remiss if I do not mention them and recognize their support. I wish to thank Daniela Axinte, Omar Al-Juburi, Benjamin Ratcliff, Tahir Paroo, Bud Vos, Teresa Tillman, Michael Burck, David Luedtke, Kathy Brewer, Brian Dawson, Michael Atkinson, and Jesse Berst.

Lastly, I wish to thank several industry veterans all of whom either influenced or supported my love and passion for this area—J. D. Hammerly, Larry Winter, Phil Crosby, Fran Shields, Jay Giri, Prof. S. S. Venkata, Tom Athay, and Kendall Demaree,

Thanks are due to the following companies, Accenture, Alstom Grid Corporation, Comverge Inc., and Calico Energy, for providing me with pictures of their offerings, which have made this book much richer than it otherwise would have been. I must add here that while I am extremely thankful for their assistance, this book does not in any way or form endorse their products, offerings, or services.

Foreword

Over nearly thirty years in the electricity industry as an executive of a global supplier of electrical operations solutions and as the CEO of an entrepreneurial venture supplying the electricity industry with innovative technology, I have seen our society's electricity dependence expand exponentially, fueled by the demands of our digital society. Every aspect of daily life depends on electricity. In fact, our society's prosperity and security hinges on the instantaneous availability and unwavering reliability of electricity and central to achieving both is the role of electric system operations. It is the role of system operations to manage the volume and flow of electricity from the generators to the customer, 24 hours a day, every day, controlling production, monitoring and configuring the grid, and responding to emergencies. More than keeping the lights on, electric systems operations spans the broad landscape of economics, information technology, and electricity consumers. Electric systems operations impacts society's behavior in response to the information it provides on cost and consumption of electricity and the impact of the electric industry on the environment.

In recent years, the public's awareness of its electricity sources increased significantly. Many people grasp the tradeoffs between the low cost, abundant, fossil-based electricity of the past and more expensive, clean, potentially intermittent, renewable resources of tomorrow. Fewer people, however, understand where electrons are generated and how they arrived at the light switch. Even fewer people have the depth of understanding of both the economics of today's electrical system and the complex technology solutions required to insure reliability and efficiency of our electricity. Dr. Subramanian Vadari is one of these rare individuals. Dr. Vadari and I have been fortunate to spend a significant portion of our careers working together to improve system operations and expanding the knowledge to allow system operations to achieve greater efficiency and reliability of electricity supply.

In over twenty years as industry colleagues, sometimes coworkers, and always friends, I have witnessed Dr. Vadari's foresight on identifying and subsequently implementing changes in technology, processes, economics, and policies enabling the electricity industry to evolve. Over his career, he maintained the unique ability to understand the specific details in the electrical system, its technology solutions, and that technology's evolution, while simultaneously grasping the implications of those details on the economics and regulatory strategy that drive electricity supply. Every change in electricity supply's technology, processes, economics, and policies, effects and likely is implemented by, system operations. Dr. Vadari is one of the few thought leaders possessing both the depth and breadth of understanding of system operations to insure our future can rely on a clean and efficient supply of electricity.

Anyone in the electricity industry having responsibility for insuring society's electricity supply, or possessing a desire to comprehend electricity delivery, must therefore understand system operations and the technology solutions system operations uses to deliver electricity with the reliability and low cost our society has come to expect. Dr. Vadari's Readers will discover a clear understanding of how system operations insures electricity gets to the light switch now and how it will continue to do so in the future.

This book provides a comprehensive review of each technology solution employed by system operations, their use, and the value delivered by their successful deployment. The importance of such an all-inclusive assessment cannot be minimized, this author expands the reader's horizons by examining both the business issues surrounding electric system operations and its human side by offering approaches for maintaining staff proficiency as system operation's complexity increases. Further, the author provides a thorough analysis of the impact on system operations from current and future investments in Smart Grid, electric vehicles, distributed energy, and emerging technological transformations. Dr. Vadari maintains a clear focus throughout on practicality, economics, and value, allowing the reader to understand not only the "what" but the "why."

J.D. Hammerly
CEO, The Glarus Group
Seattle, WA, June 2012

1

Introduction

1.1 Introduction to Utilities

A commonly held definition of a public utility is that of an enterprise that provides certain classes of services to the public including transportation, telephone and telegraph services, power, heat and light, and community facilities for water and sanitation.

- *Electric:* Includes the generation, trading, transmission, distribution, retail, metering, and customer care segments of electricity providers (see Figure 1.1).
- *Natural gas:* Includes the distribution, metering, retail, and customer care segments of end use natural gas providers.
- *Water and wastewater:* Includes the supply, treatment, distribution, metering, and customer care segments of water and wastewater providers.

The utility industry is very geographically focused compared to most other industries. Unlike a traditional product or service that can be marketed or sold anywhere in the world, utilities are constrained to provide a service only to a local region. Because of this, utilities throughout the world have evolved independently with different ownership types, product offerings, and varying degrees of regulatory oversight.

- With regards to population served, the majority of the world's utilities are owned and operated by local governments. This is especially true in

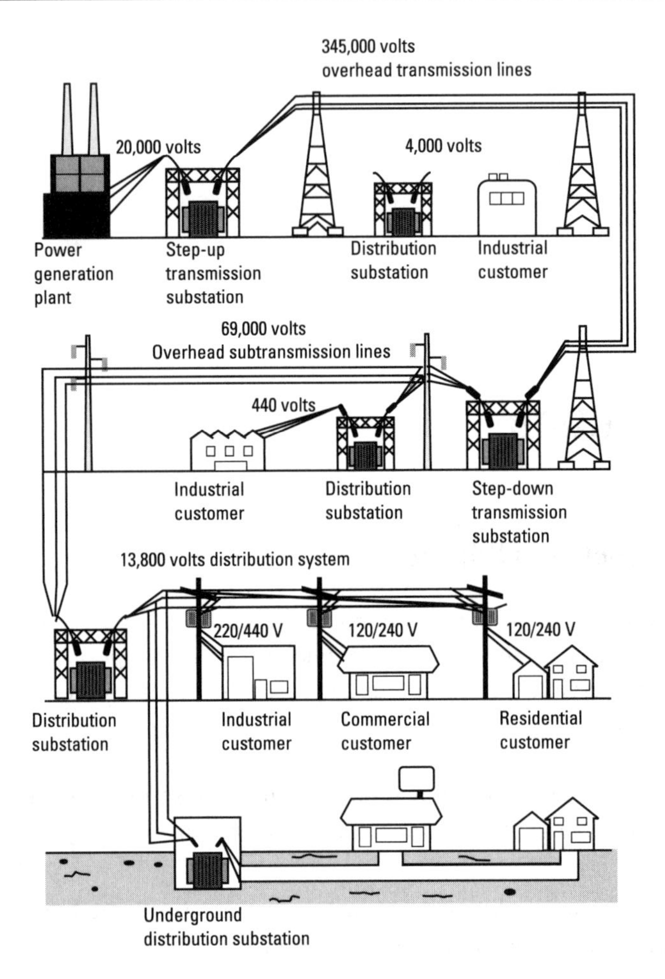

Figure 1.1 A typical utility energy delivery process.

nations of the developing world such as China and India, where a single or a small number of utilities provide services to the entire population. In Western countries, this structure varies widely by country, with many combining public ownership with a blend of governmental, private, quasi-governmental, municipal, and so forth.

- In some countries, utilities provide more than one commodity or service. Often referred to as combination or multiutilities, these entities either developed over time or were aggregated from a series of mergers and acquisitions.
- In countries where the national or local governments do not operate the utilities, the utility industries are often heavily regulated by a governmental body. In recent years, due in part to a desire to lower utility

expenses for end consumers, many electric and gas markets were opened up, allowing for competition among certain segments of the value chain. This is most widely seen in the electric industry where the generation of power and aspects of the customer supply side is often unregulated relative to the transmission and distribution segments. Due to the nature of the water industry, competition has not yet been an issue.

1.2 Explain the Electric Utility

The evolution of the electricity industry has been shaped significantly by a century of laws and regulations that govern the way electric utilities have done business (see Figure 1.2).

- By the early 1900s, the U.S. electric industry structure had evolved into vertically integrated (distribution, generation, transmission) utility companies operating under state or local regulation that reinforced the exclusive nature of the local franchise;
- Early electric utilities were inefficient and redundant in the services they provided;
- In response to this early chaos, governments developed regulations to eliminate redundancies in equipment and service costs;
- As the industry evolved, the individual owners of the early power plants gave way to investor-owned companies, which in turn evolved into larger enterprises;
- In the past, geographic lines and borders delineated most electricity markets, but thanks to recent developments and changing regulations worldwide, utilities increasingly have the capability to cross into each other's territories.

Producing, delivering, and selling electricity to end users involves a set of basic processes. The performance of these activities may have moved (in some cases) from a vertically integrated utility in which all of the functions were performed by one company to a model in which some of the functions may be performed by others within a common jurisdiction.

1.2.1 Generation

Electricity is the flow of electrical power or charge. It is a secondary energy source which means that we get it from the conversion of other sources of en-

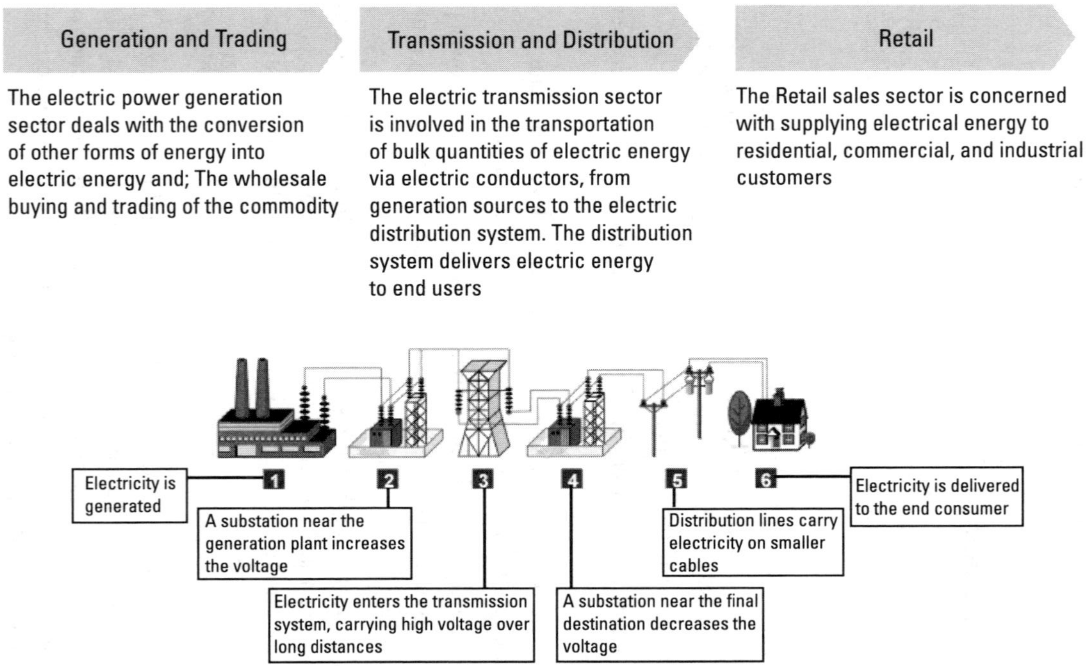

Figure 1.2 An electric utility delivery value chain.

ergy, like coal, natural gas, oil, nuclear power and other natural sources, which are called primary sources. The energy sources we use to make electricity can be renewable or nonrenewable, but electricity itself is neither renewable nor non-renewable (see Figure 1.3).

- Electric power generation is the conversion of other forms of energy into electric energy. Bulk energy is usually generated from fossil fuels (coal, natural gas, and oil), nuclear fuel, geothermal steam, falling water, and alternative and renewable energy resources.
- Rotating turbines attached to electrical generators produce most commercially available electricity. Turbines are driven by a fluid such as steam, which acts as an intermediate energy carrier.
- A conventional power station uses a turbine, engine, water wheel, or other similar machine to drive an electric generator or a device that converts mechanical or chemical energy to generate electricity. Steam turbines, internal combustion engines, gas combustion turbines, water turbines, and wind turbines are the most common methods to generate electricity. Most power plants are about 35% efficient. That means for every 100 units of energy that go into a plant, only 35 units are converted to usable electrical energy.
- Electrical energy cannot be stored economically, so it must be generated and instantaneously delivered based on customer demand. Consequently, an electric utility company must own production facilities (or procure supplies) capable of meeting the maximum demand on its system.

Essentially the classification of generation facilities/power plants can be by fuel (Figure 1.4) or by prime mover (turbine).

Classification of power plants by fuel (by primary source of energy):

- Fossil-fuel based;
- Hydroelectric;

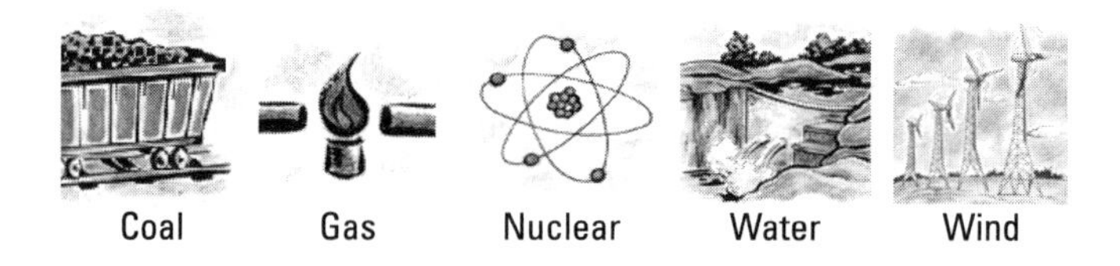

Figure 1.3 Electricity generation and its forms.

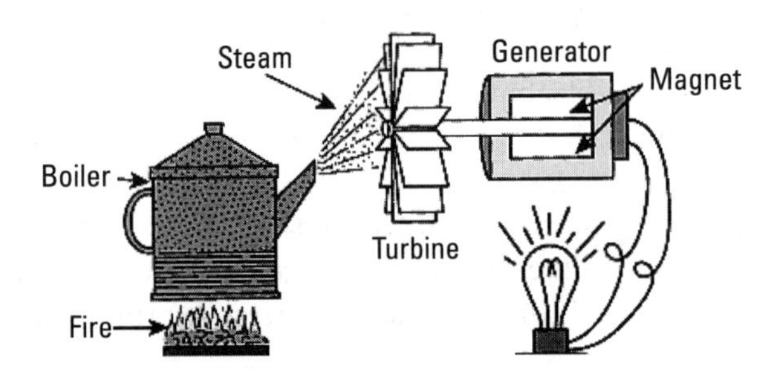

Figure 1.4 Conceptual view of a fuel-fired generator.

- Nuclear;
- Cogeneration;
- Other (solar, wind, geothermal, biomass).

Classification of power plants by prime mover (by form of energy that drives the turbine)

- Steam turbine;
- Water turbine;
- Gas turbine;
- Combined cycle;
- Other (microturbine, stirling engine, internal combustion reciprocating engine).

Generally, large power plants are somewhat centrally located in the electric grid and can be either coal-, gas-, or oil-fired, nuclear powered, hydro (large dams), and so forth. Some of the newer sources of generation also include large wind farms or solar farms.

Generation can also come from distributed sources or locations like home PVs, cogenerations (cogens) (within industrial complexes), or diesel generation sets (gensets).

After electricity is generated, it must be transmitted and distributed to the end user's specific site. The transmission and distribution system must be capable of transporting electricity, in the amount needed, to a wide variety of customers—residential, industrial, and commercial.

1.2.2 Transmission

Transmission involves the transportation of bulk quantities of electric energy via electric conductors, from generation sources to an electric distribution system, load center, or interface with a neighboring control center.

Transmission is the process of conducting the flow of electricity at high voltages from the points of generation to the locations of groups of electricity users (such as neighborhoods, industrial parks, and commercial centers).

- The electricity produced by a generator travels along cables to a transformer substation, which changes electricity from low voltage to high voltage.
- To transmit electricity effectively over long distances while minimizing power losses, utility companies use high-voltage transmission lines.

  Ohm's law states that as the transmission voltage increases, the amount of current flowing in the conductors reduces (assuming that the power transferred is still the same). Given that energy losses are proportional to the square of the current flowing (explained in more detail in Chapter 3), any reduction in current will be followed by a significant reduction in power losses. For example, if the current flow is reduced by a factor of 2 (by increasing the voltage by a multiplier of 2), the energy lost due to the flow of power is reduced by a factor of 4. (Figure 1.5).

Figure 1.5 Electric transmission system.

- Transmission lines are used to carry the electricity to a distribution substation. Distribution substations have transformers that change the high voltage electricity back into lower voltage electricity.
- Transmission lines can be supported on large poles or towers or may be underground in more urban environments.
- Placing lines underground helps to reduce outages due to weather and vegetation; however they cost many times more than above-ground systems, due primarily to high construction costs.

Engineers design transmission networks to transport the energy as efficiently as feasible, while at the same time taking into account economic factors, network safety, and redundancy. These networks use components such as power lines, cables, circuit breakers, switches and transformers, and substations, where much of this equipment resides.

The transmission system is generally distinguished from other parts of the grid based on voltage class. Generally, the voltage class of equipment above 132 kV is considered as transmission. In some locations 69 to 70 kV is also considered as transmission.

Key characteristics of transmission systems include:

- Large and high transmission towers carrying large quantities of power across long distances.
- Generally networked, meaning there are multiple paths to the deliver power to the same location. Provides for higher levels of redundancy on outages.
- Higher voltages used for transmission to reduce losses.
- Transmission equipment is considered to be of higher criticality due to the ability of key outages to spread across larger locational jurisdictions.

1.2.3 Subtransmission

Subtransmission is part of an electric power transmission system that runs at relatively lower voltages. It is uneconomical to connect all distribution substations to the high main transmission voltage, because the equipment is larger and more expensive. Typically, only larger substations connect with this high voltage. It is stepped down and sent to smaller substations in towns and neighborhoods. Subtransmission circuits are usually arranged in loops so that a single line failure does not cut off service to a large number of customers for more than a short time. While subtransmission circuits are usually carried on overhead lines, in urban areas buried cable may be used.

There is no fixed cutoff between subtransmission and transmission or subtransmission and distribution. The voltage ranges overlap somewhat. Voltages of 69, 115, and 138 kV are often used for subtransmission in North America. As power systems evolved, voltages formerly used for transmission were used for subtransmission, and subtransmission voltages became distribution voltages. Like transmission, subtransmission moves relatively large amounts of power, and like distribution, subtransmission covers an area instead of just point to point.

1.2.4 Distribution

Electricity distribution is the final stage in the delivery of electricity to end users. A distribution system's network carries electricity from the transmission or subtransmission system and delivers it to consumers. Typically, the network would include medium-voltage (less than 50 kV) power lines, substations and pole-mounted transformers, low-voltage (less than 1 kV) distribution wiring, and sometimes meters.

An electric distribution system is tasked with delivering electric energy to end users. It links the transmission system with utility customers (see Figure 1.6).

- The distribution systems begin at the substations, where power transmitted on high-voltage transmission lines is transformed to lower voltages for delivery over low-voltage lines to the consumer sites.

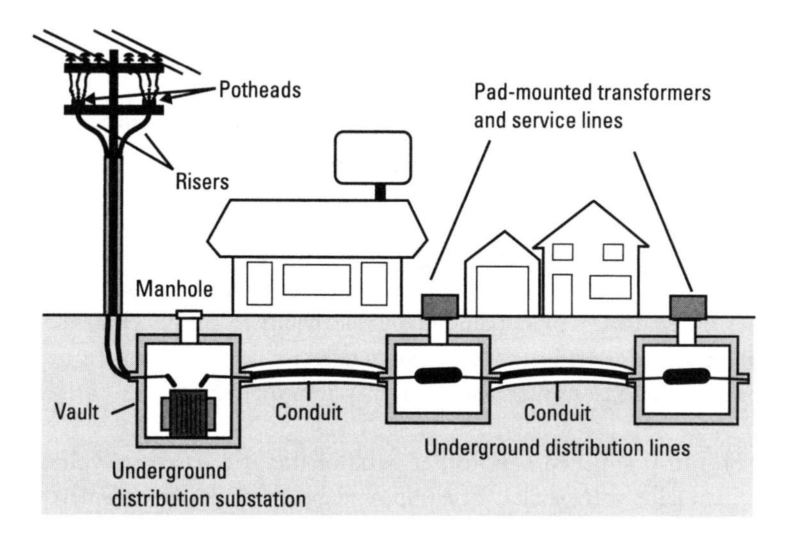

Figure 1.6 Electric distribution system.

- Distribution lines [2] may be above ground or underground depending on the geography, the weather zone, and the age of the building or community.
- At a customer's site, meters attached to the distribution lines measure the amount of electricity used during a particular period so that the utility may charge the appropriate sum to each account.
- Customers buy electricity in units called kilowatt hours (kWh). A kilowatt hour is equivalent to a 100-watt lightbulb burning for 10 hours or a 10-watt bulb burning for 100 hours.
- Distribution is considered a "natural monopoly" and is likely to remain a regulated function because duplicate systems of lines would be impractical and costly.
- A utility's field crews are responsible for maintaining the electric transmission and distribution grid through vegetation management and line repair.

The distribution network businesses are the owners of the last leg of the electrical infrastructure, from the transmission networks to the customer's meters. They collect electricity from the transmission networks and deliver it to the end users. The core business of disseminating electricity is supported by a range of field and technical services, including:

- Construction of substations, underground, and overground voltage lines;
- Erection of poles and wires to customers' premises and installation of meters;
- Regular inspection, testing, and maintenance of all equipment;
- Erection and maintenance of street lighting.

At the end of a distribution line, an electric meter is located on the customer's premises to measure the customer's usage in kilowatt hours for billing purposes. The business of actually buying electricity from the wholesale market and selling to end users is performed by retailers.

Key characteristics of a distribution system include:

- Still distinguished from other parts of the grid based on voltage class. Generally, voltage class of equipment below 33 kV is considered distribution. In some locations, 69 to 70 kV is also considered distribution.

- Smaller towers could even be wood-pole, carrying smaller quantities of power across shorter distances.
- Lower amounts of power carried and so losses are less of a concern when compared to transmission.
- Generally radial, meaning that the distribution lines go radially from the substation to the load (generally a residence or commercial or a small industry) with main feeders and tap lines.
- Both in downtown-like areas and other areas of criticality, loops (networks) are being introduced to increase reliability of power delivery.
- The distribution equipment is considered to be of lower criticality due to limited ability of key outages to spread across larger locational jurisdictions.

While transmission systems are somewhat similar in their makeup and design across most countries and jurisdictions, we find the distribution systems to demonstrate a wide variation.

1.2.4.1 Some International Differences

In many areas, delta three-phase service is common. Delta service has no distributed neutral wire and is therefore less expensive. In North America and Latin America, three-phase service is often a Y (wye), in which the neutral is directly connected to the center of the generator rotor. The neutral provides a low-resistance metallic return to the distribution transformer. Wye service is recognizable when a line has four conductors, one of which is lightly insulated. Three-phase wye service is excellent for motors and heavy power use.

Many areas in the world also use single-phase 220 or 230V residential and light industrial services. In this system, the high-voltage distribution network supplies a few substations per area, and the 230V power from each substation is directly distributed. A live (hot) wire and neutral are connected to the building from one phase of the three-phase service. Single-phase distribution is used where motor loads are small.

1.2.4.2 Americas

In the United States and parts of Canada and Latin America, split-phase service is the most common. Split-phase provides both 120 and 240V service with only three wires. The house voltages are provided by local transformers. The neutral is directly connected to the three-phase neutral. Socket voltages are only 120V, but 240V is available for heavy appliances because the two halves of a phase oppose each other.

1.2.4.3 Europe

In Europe, electricity is normally distributed for industry and domestic use by the three-phase, four-wire system. This gives a three-phase voltage of 400V and a single-phase voltage of 230V. For industrial customers, three-phase 690/400 volt is also available.

1.2.4.4 Japan

Japan has a large number of small industrial manufacturers, and therefore supplies standard low-voltage three-phase service in many suburbs. Also, Japan normally supplies residential service as two phases of a three-phase service, with a neutral. These work well for both lighting and motors.

1.2.5 Customer

As in any other market, the customer is the final end user of the product generated and distributed by the electric utility. In fully regulated environments the incumbent utility manages all aspect of power generation and delivery and also owns the customer relationship. However, in places where there is competition to supply electricity, the customer relationship or the mandate to deliver power reside with the retail energy provider [3].

However, in the United States, in most markets the cost of the actual commodity (generally the largest segment) is separated out on the customer's electricity bill whether it is supplied by a different entity or the incumbent utility.

The utility (or the retail provider) has to follow some key processes to support the energy delivery process:

- Meter-to-Cash
  - *Meter reading.* Once a meter has recorded a customer's electricity use, the data needs to be collected by the utility in order to produce a bill. This has traditionally been done manually by a meter reader viewing each customer's meter and recording the usage. While this is still the most common form of data collection, smart meters [4] are enabling utilities to collect data remotely over telecommunication lines, eliminating the need for personnel.
  - *Billing.* After the data is collected, a bill is calculated, printed, and mailed to the customer. Some utilities offer this electronically as well via Web sites or automatic bank drafting.
  - *Payment and presentment.* Once the payment is received from the customer, it is processed and credited to the utility's account.

- *Collections.* If payment is not received from the customer within a certain time frame, it is considered in arrears and various attempts are made to collect payment. If the utility is unable to collect past accounts the amounts are written off the financials.
- *Regulatory affairs.* Many utilities, especially in the European Union and the United States, are required to submit periodic reports of their operations and produce documentation for rate increase requests, permission for new generation or transmission and distribution (T&D) projects, merger and acquisition activities, and the like.
- *Customer care.* Utilities often offer customer contact centers, Web sites, walk-up payment facilities, and other programs designed to create a positive experience for their customers.

Electricity consumers are divided into classes of service or sectors (residential, commercial, industrial, and other) based on the type of service they receive:

- Single-family residential;
- Multifamily residential (condominium/apartment complexes);
- Commercial;
- Small industrial;
- Large industrial.

The voltage level coming into the customer's premises could be anywhere from 110/220 to transmission level voltages and is dependent on the type, amount of load, and the criticality for uninterrupted levels of power.

Customers (at all levels) also tend to have localized sources of power supply (gensets, etc.) to drive their business processes during power outages. This happens more in third-world countries than in others and is mainly due to lack of good reliable power service from the utility.

1.3 Electric Utilities: A U.S. Historical Perspective

1.3.1 First Came PUHCA

The Public Utility Holding Company Act of 1935 (PUHCA), was passed by the United States Congress to regulate utilities and protect investors and consumers from the economic disadvantages produced by a small number of holding companies that owned most of the nation's utilities. This act put a variety of limitations on them. A summary of limitations imposed on them included limiting their operations to a single state, subjecting them to effective state regulation, or nonutilities, such as oil companies or investment banks, to own utilities. The act also prevented utility holding companies engaged in regulated businesses from engaging in unregulated businesses.

PUHCA requires public utility holding companies to register with the Securities and Exchange Commission (SEC). The SEC has substantial and wide-ranging authority over public utility holding companies. If necessary, the SEC may control new stock issues of a PUHCA-registered holding company, prevent the buying and selling of holding company assets, and to a large extent determine the terms of the acquisition of holding company property and stock.

PUHCA resulted eventually in the reduction of public utility holding company influence. For example, holding company control of electricity generation was reduced from 75% of all generation to just 15% or so. PUHCA, by effectively reorganizing the electric and gas industries, facilitated greater federal and state regulation of utility wholesale and retail prices and conditions of service.

1.3.2 Along Came Deregulation

With the Federal Energy Regulatory Commission (FERC) Order 888/889 mandating open access of the nation's transmission network, the electric utility industry began its most significant transformation since the creation of PUHCA. Once highly regulated and organized into vertically integrated entities, the industry is being restructured into a competitive arena made up of veterans as well as new players from other industries. They are all facilitating change through the establishment of new business processes and information flow and in the development of the functions that support a deregulated marketplace.

Deregulation has resulted in major changes to the organization structure of most utilities and more importantly resulted in the creation of a new set of companies comprised of regional transmission operators/independent system operators (RTOs/ISOs) (see Figure 1.7) [6]. Within vertically integrated utilities, Chinese walls needed to be implemented to demonstrate nondiscriminatory access to transmission as well as allowing the distribution company to buy power from the most market-competitive generation in the system. As a result, even within utilities, the transmission as well as the generation (and/or trading) group needed to separate themselves from an organizational and business process perspective from the rest of the utility creating wires-only businesses.

1.3.3 Then Came Smart Grid

The electricity system in the United States received renewed attention after the August 2003 blackout that impacted more than 50 million customers across the Northeast United States and caused billions of dollars of damages to the U.S.

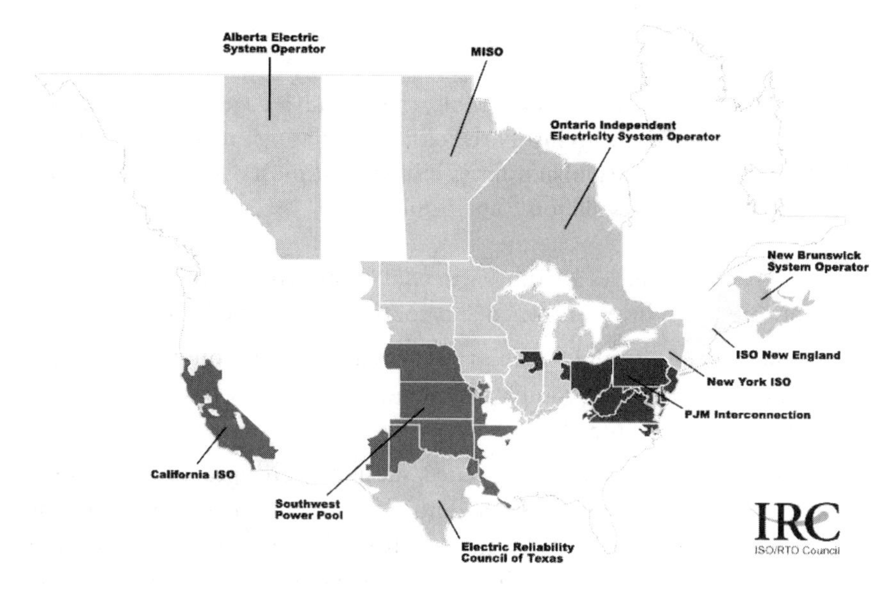

Figure 1.7 RTO/ISO Map of North America.

economy. This blackout became a catalyst to deliver a call to action as the event exposed the United States' dependency on such a vulnerable infrastructure. The call to action was that the present situation was felt to be unacceptable and change was needed to ensure enhanced safety and reliability across the system. The intelligent network is one of the results of that call to action.

1.3.4 A Global Outlook

Much of Section 1.3 has focused on the changes within the electric utility industry in the United States. However, for much of the rest of the world, electric utilities are either government owned or heavily controlled by the local or federal government in that area. With the advent of deregulation in many countries, much of this has changed with clearer delineation between wholesale and retail services. In many of these countries (United Kingdom, Australia, the Netherlands, etc.), the distribution company is completely private. It is not the intent of this book to go into too much detail on the corporate structures of every country.

1.4 Utilities and Regulation

Utility industry regulation has historically been the means for local and national governments to reap the benefits of these natural monopolies—benefits such as cost advantages and essential energy service—without suffering the consequences of uncontrolled prices.

In order for regulation to be feasible, the cost advantage of the natural monopoly must be sufficient to justify the cost of regulating the industry (this has been the belief about the global utilities industry for most of the 20th century).

Key U.S. utility regulation facts include:

- All types of utilities are subject to regulations whether they are public or private (both investor owned and cooperatively owned) utilities.
- There are more public than private gas and electric companies, but private gas and electric companies provide the majority of the energy consumed in the United States.
- Although the industry's regulated status has changed in recent years, for the majority of the 20th century U.S. gas and electric industries were fully regulated natural monopolies, with protected franchise areas, regulated rates, and the obligation to serve the customers in their areas.
- Regulation of interstate transportation of gas and electricity falls under national jurisdiction, while distribution to individual customers is generally a state responsibility.

Utility industry regulation is at the national or local level, each having its own jurisdictional responsibilities. Oversight responsibilities of regulatory agencies include:

- Approving energy rates;
- Monitoring competition and access;
- Enforcing national or local energy laws;
- Monitoring safety and environmental issues;
- Monitoring system reliability;
- Ensuring that franchise and obligation-to-serve requirements are met.

To maintain reasonable rates of return for utility companies, public utility commissions have established procedures for setting and changing rates. The process of setting rates includes the following:

- In order to obtain permission to adjust base rates, a utility company must submit a rate case to the local public utility commission;

- The commission reviews the rate case documentation and considers the points of view expressed during a public hearing and makes a ruling as to the rate adjustment requested by the utility;
- A utility company may or may not attain its expected rate of return regardless of approved rates, because other factors, such as mild weather or unexpectedly high operating costs may significantly affect the bottom line.

At a more general level, utilities are one of the most heavily regulated services in most countries with processes somewhat similar to that in the United States. However, most of these countries do not have the patchwork of state and federal regulatory mechanisms that exist in the United States.

Endnotes

[1] Distribution lines when they leave the substation are also called feeders or tap lines, depending on where they are in the system. Generally, feeders leave the substation and tap lines tap off the feeders to deliver power to the premise.

[2] A retail energy provider (REP) is the nomenclature used in the Texas deregulated market.

[3] A smart meter is usually an electrical meter that records consumption of electric energy in intervals of an hour or less and communicates that information at least daily back to the utility for monitoring and billing purposes. Some smart meters enable two-way communication between the meter and the central system and can also gather data for remote reporting. Such an advanced metering infrastructure (AMI) differs from traditional automatic meter reading (AMR) in that it enables two-way communications with the meter.

[4] Independent system operators grew out of Order Nos. 888/889 where the Commission suggested the concept of an independent system operator as one way for existing tight power pools to satisfy the requirement of providing nondiscriminatory access to transmission. Subsequently, in Order No. 2000, the Commission encouraged the voluntary formation of regional transmission organizations to administer the transmission grid on a regional basis throughout North America (including Canada). Order No. 2000 delineated 12 characteristics and functions that an entity must satisfy in order to become a regional transmission organization.

2

Define System Operations

The wide distribution of electrical power in the 20th century brought light to the world and power to almost every pursuit and enterprise in modern society. Consider its impact on everyday life—lighting, heating and air conditioning, refrigeration, computers, transportation, communications, medical technologies, food production—the list is endless. Several key engineering innovations made this possible, including the turbine generator, the use of alternating current (AC), techniques to obtain electrical power from various resources (fossil fuels, water, sunlight, nuclear), and the construction and refinement of massive transmission systems. Electrification is responsible for innumerable developments that have made life safer, healthier, and more convenient; so much so that it is hard to imagine our lives without it. It runs the smallest electric appliances in homes and offices, the mammoth computers that control power grids and telecommunications systems, and the machinery that produces consumer goods. Its impact is vast, and it has touched the majority of people on the planet [1].

Every day, with clockwork precision at every utility, different generators (some hydro, some thermal, some gas, some by other means) transform other forms of energy into consumable electricity. Just as surely, a backbone of transmission lines then delivers this power to the power-hungry centers of consumption: lighting lights, driving industries, powering office buildings, and bringing life to various homes all across the land. The electric utility's system operations centers are responsible for safeguarding these assets by providing reliable electric power to its customers.

Electric system operations is the entity that is responsible within a utility for managing the transmission and distribution grid. In the past, this role was purely technical and taken on by a grid operations group within the vertically integrated monopoly utility. A typical responsibility was balancing transmission line energy flows to ensure network stability. In the future, this team will need significantly new business capabilities and will be faced with a much more complex operational role. On the business side, they will need to perform contract management, develop and maintain customer relationships, and reconcile physical energy consumption through financial settlements. On the operational side, they will need to coordinate scheduling and manage network congestion for an increasingly complex environment where industrials, power marketers, and generators alike will all be looking for the best deal.

2.1 System Operations

Like something from a science fiction movie, the electrical grid is a large, complex machine. Nations and economies depend on it and consumers just expect it to work. Against these demands, electricity providers must ensure that all of the equipment and processes operate together, safely and reliably.

So where does system operations fit into a utility's overall ability to meet customer needs today? The short answer is "everywhere." Asset management is about deploying the right assets at the right time. Work and resource management ensure that you get the right work done, at the right cost. And system operations has traditionally focused on operating the network to ensure the right levels of safety, reliability, and efficiency in everything that relates to real-time grid management.

While other areas in utilities can be as much as 10 to 15 times larger in terms of budget and personnel, system operations is still a significant driving force behind energy delivery and customer service. It can also be a bottleneck because even if utilities make changes in other critical areas such as asset management, system operations must support such initiatives for them to work effectively (e.g., helping work crews to locate and service outages as well as coordinate work safely and cost-effectively).

Market forces are accelerating change in utilities: the workforce is aging, skills are becoming more scarce, government and advocacy groups are imposing new regulations (and more fines), competition is changing, and the customer looms ever more important before an industry that, in many cases, is simply trying to keep up. Today's electrical environment only adds to the complexity of the system operator's job, which requires more work with the same or fewer resources.

Electric system operations in today's environment is not the same as it was a generation ago. There are many new challenges utilities face in the modern environment from both internal pressures to provide efficiency and performance gains and outside pressures that are looking for better service and reliability.

To summarize, the electric system operations capability is expected to operate the electric T&D network safely and efficiently.

2.2 Key Drivers for Systems Operations

There are a common set of drivers that impact electric system operations in a critical manner, driving the need to innovate as well as the need for increased efficiencies (see Figure 2.1). They are all focused around a need to demonstrate:

- Transparent operations;
- Rapid restoration from outages;
- Ability to monitor and operate the system reliably;

Do the above steps efficiently.

2.2.1 Impact of Drivers on Distribution

The inability to restore outages in a faster manner coupled with an aging workforce and aging infrastructure continues to be one of the main issues faced by

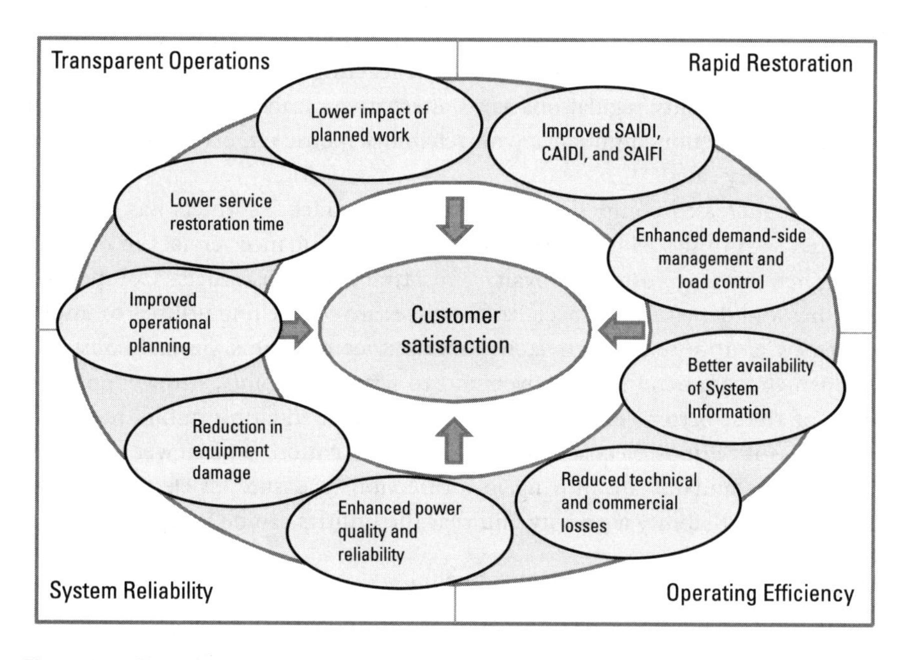

Figure 2.1 Key drivers and factors impacting system operations.

distribution utilities. An ongoing strong focus on industry-wide performance metrics like the system average interruption duration index (SAIDI), the customer average interruption duration index (CAIDI), and the system average interruption frequency index (SAIFI) [2] have made distribution operation a business platform with rising costs and unsatisfied customers potentially seeking services from other utilities. Control center consolidation, implementation of smart grid and smart meter, extended integration of key systems beyond AMI and supervisory control and data acquisition (SCADA) [3] are some of the solutions utilized by the utilities, but views have been mixed due to lower than expected performance results. The proper use of a distribution management system (DMS) can help to enable the utility to realize full benefits as various programs are implemented to deal with their major operating issues.

2.2.1.1 Increased Competition

Utilities are facing significant external pressures that are driving a business model change for the industry from a customer perspective. States have been increasingly active in promoting competition and market restructuring since the approval of the Energy Policy Act of 1992 [4]. Competition is driven from both a retail perspective and by the threat of municipalization. In both scenarios, the result is the need for the utility to decrease price and increase customer satisfaction to retain customers. State regulators and their constituents no longer tolerate substandard service and high prices.

Additionally, advocacy and industry watchdog groups have been highlighting the inadequate service provided by electric utilities, and consequently state public utility commissions have been reacting by imposing increasingly demanding reliability regulations as an alternative means to increase customer satisfaction. The noncompliance with reliability metric targets is often coupled with monetary fines.

The increased competition to retain and attract customers has escalated the need for utilities to improve performance and enhance services in order to strengthen existing customer loyalty and attract new customers. Competitive pressures and demands for specialized services are compelling utilities to implement new and more innovative customer service programs. Several consumer groups are recommending improvements to service reliability, stronger enforcement of rules, better consumer protection, and mandating public reporting processes for various electric utilities across the nation. The newer customer satisfaction standards are moving in a direction to ensure all electric utilities make service reliability a priority and that the utilities can be held accountable.

2.2.1.2 Load Growth and Aging Infrastructure

Utilities are witnessing rapid expansion of power distribution systems in size and complexity coupled with a strict limitation of investment resources that severely limits the growth and improvement of the aging infrastructure. Not only are there increased customers in a jurisdiction for an operator to manage, each customer is now putting more load onto the system with a multitude of electronic gadgets (such as computers, televisions, and appliances) amassed in a typical home or business. These additional gadgets have a nonlinear operating characteristic which, in turn, impacts the power quality in the network. In addition, this also results in the capacity on the distribution network getting more constrained putting increased pressure on the system during an outage or an emergency situation.

In addition, operators must manage various equipment types in which many can be decades old and each with their own range of device ratings, trip settings, and normal/abnormal status. An operator must either remember each equipment's rating (unlikely) or manually look up this information for each device on the circuit that is involved in a switching process. This puts additional pressure on the distribution operator to find ways to transfer load, or in the worst case, shed load.

Historically, the primary approach a utility would employ to mitigate the effect of load growth and aging infrastructure would be to spend significant dollars to upgrade the electric system. In many cases, the investment could have been delayed and the life of the infrastructure in place could have been extended by the more efficient operation of the grid.

2.2.1.3 Data Overload

In today's technology-driven environment, workers often have an overload of data forced on them, and a distribution system operator is no exception. The distribution system operator of today manages multiple applications to complete their day-to-day responsibilities. Many utilities have separate user interfaces for operating SCADA devices, managing outages, interacting with field resources, entering outage information for customer communications, and logging and writing switching orders. In storm situations, operators find themselves managing multiple outages, toggling back and forth, trying to keep track of it all. Many times operators find themselves duplicating work effort and entering the same information multiple times into multiple systems. All the while, the operator is continuing to receive telemetry data from the field, which he or she has not had time to assess.

A distribution system operator is simply overloaded with data and verbal commentary. The data comes from different user interfaces and the verbal commentary from field personnel. The operator does not have the time to convert the imputs into useful information that can be leveraged to make decisions. Without the time or tools to make sense of all the inputs, much of it is wasted, leading to an increased potential for less than optimal operation of the distribution system.

2.2.1.4 Aging Workforce

The aging workforce has and will continue to be a significant issue in the utility industry. In the next 10 years, the utility industry expects a significant reduction in its workforce as the average age of a utility worker is 50 years [5]. For some utilities, the number of expected retirements is over 40% of its current staff, including supervisory personnel. This departure of key personnel represents a loss of knowledge capital, experience, and capability and results in a huge risk to effectively and efficiently operate the distribution network.

Within distribution system operations, the long lead time required to onboard a new hire makes this even more pressing. Depending on the utility and the jurisdiction, the lead time can range from 1 to 2 years to be capable of operating a network. Developing the proficiency and knowledge of the "traps" within the jurisdiction takes an even longer time. All of this intensifies the criticality of the risk to the utility.

If adequate response to this is not planned, as veteran operators retire, the utility could be left with a lack of resources and a lack of knowledge and skill to manage the grid safely and efficiently.

2.2.1.5 Desire to Reduce Outage Duration and Frequency

Utilities are always working to reduce the frequency and duration of outages. With each outage their customer base is less satisfied with the electricity service and there have been many cases of cities breaking away from their utility. In addition, SAIDI and SAIFI costs are impacting revenues as utilities are paying millions in public utility commission (PUC) [6] penalties.

2.2.1.6 Need for Increased Visibility

The manner in which a utility responds during an emergency is critical not only for safety but also for customer satisfaction. Paper wall maps are difficult to use to pinpoint the location of outages and know the state of equipment and location of field crews. After many phone calls and sectionalizing circuits more knowledge about the outage is gained but unfortunately only the single control center has the information because efficient dissemination of data could not be done electronically. Emergency centers are not notified in a quick enough manner. In addition, customers need to know the state of the emergency as the system is restored. Without an efficient real-time system, customers are not updated on the situation, causing them more angst.

2.2.1.7 Reduction of Risk from Disasters

In the current paper wall map environment, there is no backup map in case of a regional disaster. Today colored pushpins are inserted onto a map of the network to indicate open or closed switches and crew locations. Tagging is done with markings on a paper hooked onto the location to indicate a location where a crew is dispatched. There have been cases where pins and tags fell off the paper map during earthquakes. In this scenario there is no other backup map to indicate the "as-operated" state of the network. Control centers will spend countless hours calling crews and analyzing switching plans to recover from a disaster situation. If the disaster is more extreme and knocks out a control center completely, there is no way to transfer control of the paper map to another control center. Cross-jurisdictional operations is not possible with paper maps.

2.2.1.8 Increased Pressure to Reduce Costs

Just like any other business in these difficult economic times, utilities are also facing increased pressure to reduce costs—and this pressure is coming from all directions, including regulators, shareholders, and customers. To enable this cost reduction, utilities are looking at opportunities by consolidating control centers, shutting some of them down at night, bringing more automation into the control center, and trying to perform the same amount of work with fewer operators.

2.2.2 Impact of Drivers on Transmission

Unlike distribution, which has anticipated a transformation coming in terms of major modifications made on the system after decades of neglect, transmission has always been in a state of flux. The blackout of 2003 brought to bear areas of weaknesses, several of which were focused on the transmission system. As a result, even though much of the drivers in distribution are still relevant to transmission, there are a few new drivers that impact transmission only.

At the transmission level, disturbances have the potential to travel long distances at incredibly fast speeds, causing havoc and equipment tripping across the interconnected network. This was evident in the 2003 blackout, when the incident that originated in Ohio was able to spread across the entire eastern interconnect, leading to a blackout that impacted more than 50 million customers. In addition, the U.S. Department of Energy sees the following key points driving transmission.

Synchrophasor Implementations

A phasor is a complex number that represents both the magnitude and phase angle of the sine waves found in electricity. Phasor measurements that occur at the same time are called synchrophasors, as are the PMU devices that allow their measurement. In typical applications, phasor measurement units are sampled from widely dispersed locations in the power system network and synchronized from the common time source of a Global Positioning System (GPS) radio clock. Synchrophasor technology provides an accurate tool for system operators and planners to measure the state of the electrical system and manage power quality. Synchrophasors measure voltages and currents at diverse locations on a power grid and can output accurately time-stamped voltage and current phasors. Because these phasors are truly synchronized, synchronized comparison of two quantities is possible in real time. These comparisons can be used to assess system conditions more accurately and quickly.

Widespread implementation of PMUs in the western interconnect as well as in the eastern interconnect is bringing new levels of visibility into the transmission grid.

Power System Visualization and Situational Awareness

PMUs and a variety of sensors in the power system transmission network are bringing tremendous quantities of data into the control center. With this data comes the need to provide actionable information for faster and more confident real-time decision making by offering a unified view across the entire grid interconnection.

With advanced visualization comes the ability to provide a wider array of situational awareness (SA) capabilities to operators who are nowadays operating a more complex grid. *SA refers to the capability of efficiently gathering relevant information, processing this information to comprehend the current state, and finally, projecting the future status.* Within the utilities industry, lack of SA has been identified as a major factor behind massive outages such as the blackouts in the United States in 2003 and in Europe in 2006. It allows the operator to predict things like imminent collapses before they happen and to take actions to prevent the collapse of the system. The Eastern Interconnect Phase Implementation (EIPI) Project is one of the major projects focused on improving SA for utilities.

Figure 2.2 overlays the concept of situational awareness on a typical transmission one-line diagram. This is a fictitious power system and does not represent any specific utility power system. In this picture, the words in bold are transmission substations, the lines connecting them are transmission lines, and the arrows on them provide the direction of power flow. The highlighted areas around the Parkhill and Ceylon stations indicate violations that have been identified. There is also a new area of violation forming around the Richview station. The intent of the highlighted areas is to focus the operator's attention on the problem that is just forming and that needs their attention. It is possible that if the problem is not alleviated soon, it may escalate to something more severe, taking the grid to an unstable situation, at which point it may not be possible to bring it back to stability.

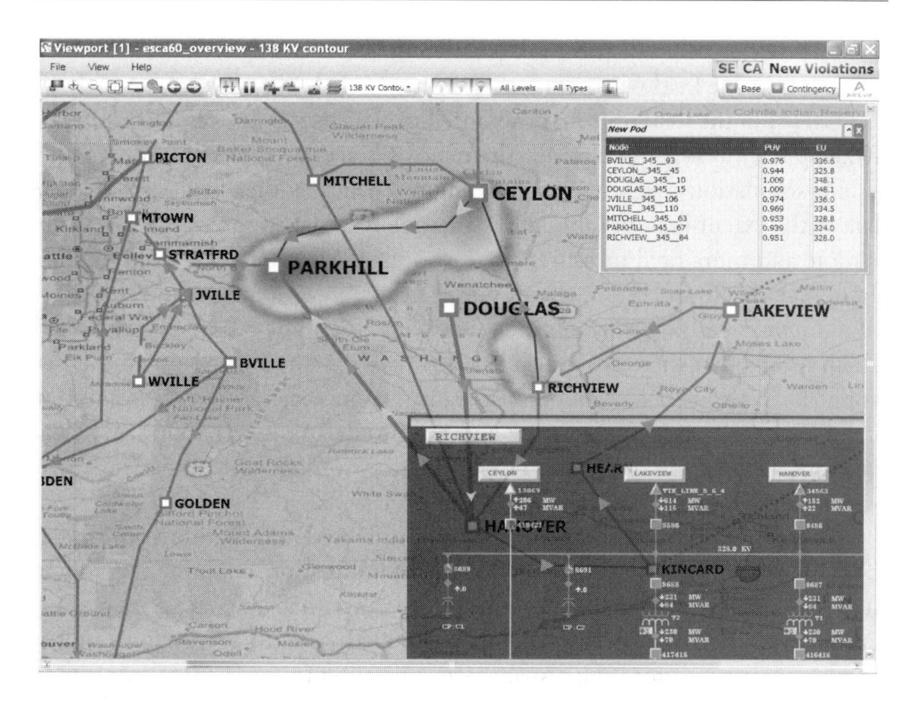

Figure 2.2 Visualization and wide-area monitoring. (From e-Terra Vision, Courtesy of Alstom Grid.)

2.3 What Changes from Transmission to Distribution System Operations?

While the term "grid operations" applies mainly to transmission operations, distribution is slowly moving out of call centers and emergency trouble-call management centers towards more formalized distribution operations centers. This is not just a simple extension of the traditional EMS/SCADA systems to include distribution systems—numerous attempts at doing this have failed and only nowadays, with the advent of DMSs that have been rearchitected from the bottom up, have resulted in success. This is because there are some fundamental differences between the two systems that then impact how they need to be managed.

Level of observability. In transmission, there are generally more measurements than observable states [7]. As a result, we have an application called the state estimator. SCADA data comes in a somewhat raw state and is then fed into the state estimator, which estimates the states of every node in the network system. The estimator uses the connected network model of the power system to develop a more accurate understanding of the system state and the rest of the EMS uses the state estimator output from then on.

The situation in distribution is different. There are considerably less measurements in the field whether coming from SCADA-like remote terminal unit (RTUs) [8] in the field or from other intelligent electrical devices (IEDs) [9]. To make up for this lack of field measurements, systems need to start with a power system simulation as the underlying engine, which is then fed with SCADA data to the extent it is available. This situation is expected to improve over time as more and more field measurements are added. Measurements from the field could include any voltage, current, power factor, MW, MVARs, and so on.

Connectivity model. Transmission systems tend to be extremely networked, which means that there are multiple paths between any two points in the network. This is done by design because transmission systems drive large amounts of power across a larger geography and need to deliver power to multiple load locations at the same time. As a result, transmission networks experience two-way power flow as a basic feature (see Figure 1.3).

Distribution networks generally tend to be radial in nature even though loops (networked connections) are slowly becoming more common than before mainly in downtown/urban areas where large power delivery requirements exist and there is a need for greater reliability. In radial systems, power flow tends to be mainly one-way from the distribution substation to the eventual consumer.

Whether it is one-way or two-way, this specific aspect of power flow has a huge impact on a large number of utility processes, many of which are associated with safety, crew-line management, and so forth. In addition, it also has

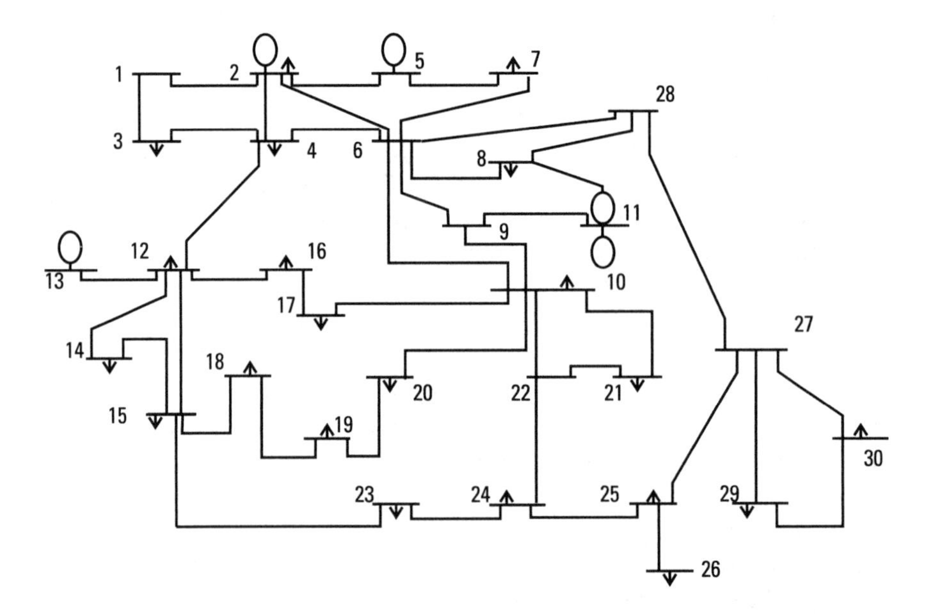

Figure 2.3 An example of a networked system: The IEEE 30Bus power system model.

a significant impact on outage management given that in distribution, there is only one way to get power to a specific location in a radial network (Figure 2.4).

Component location. With the exception of lines and their poles (or in the case of transmission towers), most components in a transmission system exist in a substation. Much of the system between substations in transmission is basically transmission lines and towers. In distribution, while there are still distribution substations, much of the components in a distribution system exist outside the substation. Components defined in the previous paragraph include transformers, circuit breakers, fuses, and relays.

The key impact of this issue is that from a system operations perspective, it is easier to get communications access to a few substations (whether that is in transmission or distribution) as opposed to providing communications access to the entire length of every line all the way to the consumer's premise. The extent of the distribution system is so much that providing ubiquitous

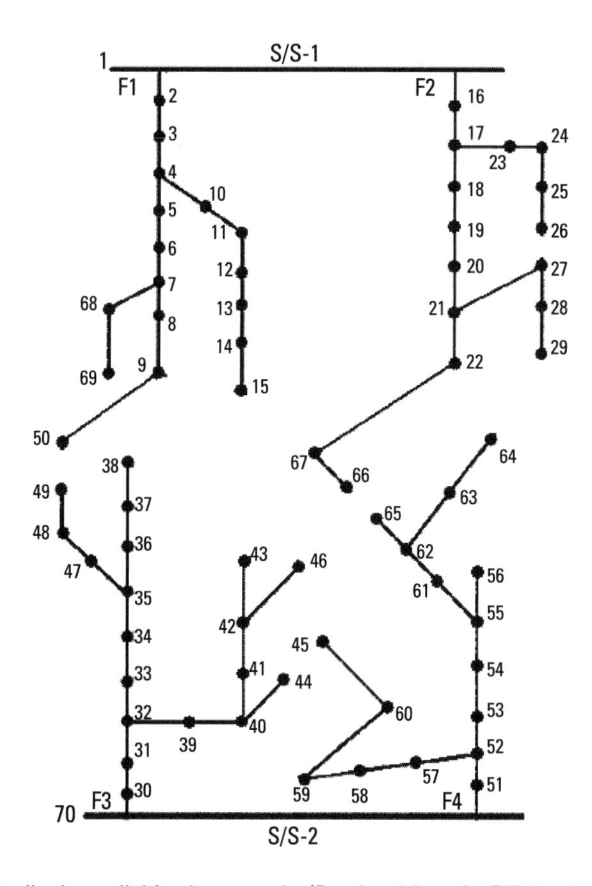

Figure 2.4 Distribution radial feeder example. (Reprinted from An Efficient Hybrid Evolutionary Algorithm Based on PSO and HBMO Algorithms for Multi-Objective Distribution Feeder Reconfiguration, Taher Niknam. © 2009 with permission from Elsevier.)

communication across the entire extent of the distribution system has been a key constraint to bringing more control and observability to a centralized location.

Three-phase versus single-phase [10]. Transmission systems tend to generally function in a three-phase balanced situation. This means that most loads tend to be connected at the three-phase level and so the three phases tend to be quite balanced. From a system operations perspective, most of the modeling is a one-line mode, meaning they tend to treat all three phases [11] as a single circuit. This allows a certain amount of simplification in the modeling.

The situation is dramatically different in distribution systems. Here the primary customer (mostly residential and small commercial) tends to consume power in a single-phase system. This means that each of the three phases could be delivering power to different consumers, causing the systems to be quite unbalanced—meaning that there is a different amount of power flow in each phase. As a result of this from a system operations perspective, distribution applications and mechanisms of control need to look at each phase separately, thereby requiring three-phase modeling supported by a three-phase unbalanced power flow solution mechanism.

New technologies and integration points. The biggest set of changes coming into the transmission system includes the installation of phasor measurement units (PMUs) and the associated applications of a wide area monitoring system (WAMS) and the advent of large wind and solar farms getting connected into the grid.

Distribution, on the other hand, has seen a tremendous amount of changes including smart meter placements, distributed generation, storage, distributed renewable, electric vehicles, and a large number of new devices and controls coming on to the grid. All of these are putting a tremendous amount of strain on the grid, making the distribution operator's life more difficult with more things to observe and control.

2.4 Distribution System Operations: An Introduction

These dramatic changes that are impacting distribution systems all have a serious impact on distribution operations, leading to the evolution of new systems such as the outage management system (OMS) and DMS. OMS, which had its humble beginnings in trouble-call management systems, has become more and more sophisticated, resulting in systems that are fully integrated with the Geospatial Information System (GIS) to provide the graphics and the underlying data and connectivity model, and SCADA to drive operator actions (see Figure 2.5).

2.5 Key Challenges Facing System Operations

System operations have seen tremendous changes from the onslaught caused by deregulation (mostly in transmission) and smart grid (mostly in distribution). These changes have resulted in both the influx of new technologies as well as changes to business processes and people-change management. These new systems have also resulted in more integration between the various operational systems (even between transmission and distribution) and the back office.

Figure 2.5 A substation. (From http://www.flickr.com/photos/palemoontwilight/3927936717/. Used with permission from Tapioca Twilight Media.)

Table 2.1
Key Problems Facing System Operations

	Operational: Management and Control	**Financial: Support the Market**
Transmission	Introduction of large wind farms directly into the transmission system have greater potential to cause instability when wind begins to either start/end blowing.	Operational data being made available to support settlement operations and dispute resolution. The challenge here is that operational data is not designed to be of revenue quality.
	Increased levels of reliance on interconnection leading to increased possibility of a problem in one area rapidly spreading to other areas (e.g., 2003 blackout). The main response to this is the introduction of synchrophasor measurement unit (or PMU). Using PMU output, one can now use WAMSs, which have the ability to analyze large amounts of data across large geographic areas and even predict the extent of damage ahead of time. The challenge now is to use this new source of data and analysis to drive new solutions to the problem.	
	Increased amounts of data coming (e.g., PMUs, as described above), make it difficult for operators to perform their monitoring and control using existing EMS display mechanisms. New methods of visualizing existing data needed along with specific solutions in the area of situational awareness.	
Distribution	Not enough visibility into the grid. Much dependence still on human intervention	Advent of AMI into distribution could allow states/utilities to support retail markets in their jurisdictions.
	Influx of new loads like EVs can be a significant strain on the system if they reach scale.	
	Distributed generation (mostly solar and some wind) creating two-way power flow in a system designed for primarily one-way power flow.	
	Large quantities of AMI/smart meters coming in to the grid, causing utilities to receive too much data that they do not know what to do with.	
	New applications like IVVC* and FLISR† are being brought into the control center to bring a greater level of sophistication and control‡ to the center.	

Both	*Cybersecurity.* The plethora of new devices getting connected to the grid and having the ability to either connect or disconnect parts of the system (or loads) creates more possibilities for cybersecurity breaches. One of the main challenges in protecting these networks is the fact that these systems were not necessarily designed with cybersecurity in mind. Rather, the security solutions have been layered on in a piecemeal fashion after the networks were operational, leaving potential room for attackers to compromise their functionality *Privacy.* Until now, the only interaction between the utility and the customer was limited to meter readers reading meters once a month (or so) and the bill. The advent of smart meters have created a situation in which the utility now not only reads the meters remotely but also knows the consumption within the premise on a 15-minute (or less) basis. This aspect, along with the introduction of home energy managers, leads to the first major insight into customer consumption patterns that could also amount to an invasion of privacy. NIST is working on developing policies for this. *Interoperability.* Interoperability is about to become a significant challenge within the smart grid community. This is because of the large number of new vendors who are entering this nascent market and making new and innovative solutions. From a utility perspective, they need all of these devices and systems to work together and seamlessly. NIST§ has been given the mandate to develop a set of standards that will allow the devices and systems to be interoperable.

IVVC = integrated volt-VAR control; FLISR = fault location identification and service restoration; NIST = National Institute of Standards and Technology.

* Integrated volt/var control (IVVC). IVVC continuously analyzes and controls load tap changers (LTCs), capacitor banks, and voltage regulators to manage system power factor and voltage. This allows utilities to flatten each feeder's voltage profile and to lower average voltages. It often results in significant energy savings while simultaneously maintaining unity power factor to eliminate technical losses.

†Fault location, isolation, and service restoration (FLISR) software can automatically sense trips (faults) in switches that are monitored and controlled by a SCADA, identify the faulted section, isolate the fault, and restore power to customers by automatically switching them to nonfaulted sections of the line. FLISR does not "fix" the problem. Utilities must continue to send crews to the site of the fault, where they verify it and make permanent repairs. But customers experience only a brief, momentary outage.

‡Several terms like feeders, capacitors, and so on are all explained in Chapter 3.

§ Under the Energy Independence and Security Act (EISA) of 2007, the National Institute of Standards and Technology (NIST) has "primary responsibility to coordinate development of a framework that includes protocols and model standards for information management to achieve interoperability of smart grid devices and systems...." To carry out its EISA-assigned responsibilities, NIST devised a three-phase plan to rapidly establish an initial set of standards, while providing a robust process for continued development and implementation of standards as needs and opportunities arise and as technology advances.

Endnotes

[1] "The Greatest Engineering Achievements of the 20th Century," http://www.nationalacademies.org/greatachievements/List.PDF.

[2] Industry-standard metrics to measure distribution grid performance. Actual definitions provided later.

[3] SCADA stands for supervisory control and data acquisition, a real-time system used to get measurements from the field into the control center systems like EMS, DMS, and so on and also to send controls to the field.

[4] Source: http://www.eia.doe.gov/cneaf/electricity/epav1/state.html.

[5] Source: http://www.eei.org/magazine/editorial_content/nonav_stories/2005-09-01-Aging.htm.

[6] PUC stands for the public utilities commission, the state-level regulatory body that regulates the utility. Generally their focus is on distribution only. They may go by different names in different states.

[7] The state of the power system is defined by the combined values of the voltage magnitude and phase angle (see information on phasors). Knowing the state of the network at every node can allow us to calculate every other variable in the network.

[8] An RTU is an electronic device that is controlled by a microprocessor. The device interfaces with physical objects to a DCS or SCADA system by transmitting telemetry data to the system.

[9] IED is a term used in the electric power industry to describe microprocessor-based controllers of power system equipment, such as circuit breakers, transformers, and capacitor banks.

[10] The more detailed concepts of three-phase versus single-phase will be described in a separate chapter later in this book.

3

Introduction to Power Systems

In this chapter, we provide a brief discussion on a set of power system fundamentals that are important to the proper understanding of how a utility grid actually operates/works.

3.1 Basic Electric Components

Electric components that form a part of the grid are responsible for the way the grid behaves under different conditions. Some of them can also be used to influence the flow of power in the transmission and distribution lines.

3.1.1 Capacitors and Reactors

In AC power systems, which cover much of today's power grid and the delivery mechanism, there are three components to load:

1. *Resistance.* A pure load (e.g., a regular incandescent light bulb) is a resistive load.
2. *Inductance.* Most loads in AC power grids are inductive in nature. Examples of devices that have significant electric power loads are refrigerators, air conditioners, and industrial motors. Most of these are considered inductive loads. Inductive loads are considered to be consumers of reactive power. When a load is called an inductive load, it means that the power consumed has two major components—real and reactive. This kind of a load also tends to bring the voltage profile of the system down. When there are a lot of inductive loads on a

feeder, the voltage tends to drop as it goes farther away from the substation.

3. *Capacitance.* While there aren't too many examples of capacitive loads in a typical power grid, capacitors are generators of reactive power. As a result, they are installed to deliver reactive power to the grid. Examples are power supplies and radio tuning circuits. Given that inductors can cause the voltage in a feeder to drop, capacitors are installed in the system to raise the voltage profile on the feeder.

Most loads on a grid have a combination of resistance, inductance, and capacitance.

3.1.2 Transformers

The transformer is based on two principles: first, that an electric current can produce a magnetic field (electromagnetism), and second, that a changing magnetic field within a coil of wire induces a voltage across the ends of the coil (electromagnetic induction). Changing the current in the primary coil changes the magnetic flux that is developed. The changing magnetic flux induces a voltage in the secondary coil.

Figure 3.1 shows an ideal transformer. In this device, there are two sets of coils: the primary and secondary, which are wrapped around a core of a magnetic material. When alternating current (AC current) passes through the primary coil, it creates a magnetic field that flows around the core. When this field interacts with the secondary coil, it creates a voltage on that coil. The difference in the voltage between the primary and the secondary coils is defined by the difference in the number of the windings around the core. If a load is connected on the secondary coil (as shown in Figure 3.1), current will also flow in the secondary circuit.

The transformer is a fundamental component of the AC-power-based power grid. The transformer allows the voltage on the grid to be increased or decreased as needed. If we move along the utility value chain, the generator generates at a certain voltage. This is then stepped up using a transformer to transmission level voltages and transmitted across long distances. Once it comes toward a load center, another transformer steps down the voltage. As it gets closer to the actual load consumption, a pad-mounted transformer further steps down the voltage to the level that allows it to be consumed inside the house.

Losses in a power line are directly proportional to the square of the current flowing in the line. So, when power is transmitted over long distances, the voltage is stepped up to higher levels to reduce the current flowing through the line and as a result reduce the losses. As we get closer to the point of consump-

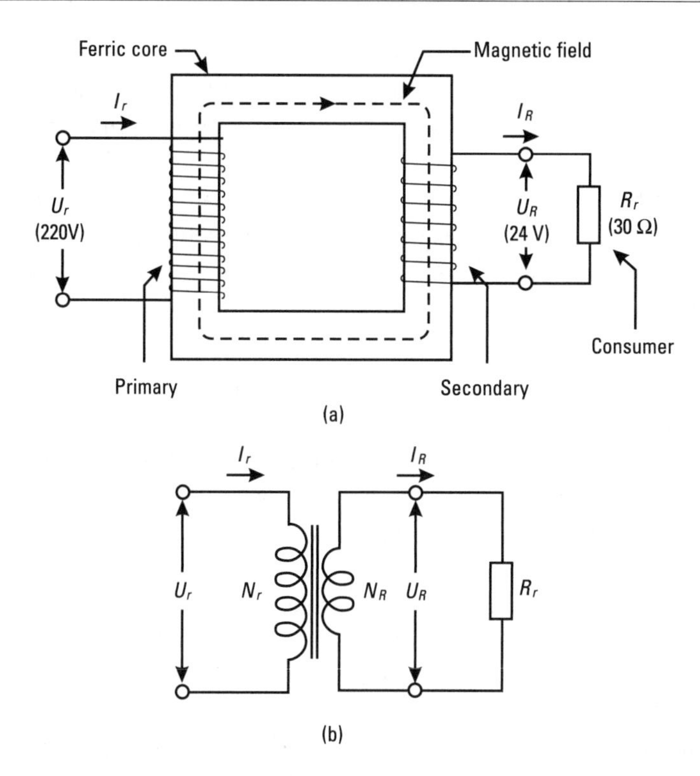

Figure 3.1 Typical transformer internals.

tion, step-down transformers allow the voltage to be brought down to a level that allows the consumption to take place.

This ability for the voltage to be stepped up and down is a major discriminating factor between AC power and DC power. While newer devices, like choppers and boosters, do this quite effectively using power electronics. The problem with DC is the ability to break the large currents because there are no zero crossings.

3.1.3 Switches

Switches, as the name indicates, are devices that either stop or allow the flow of power. In an electric circuit, we can see several types of switches.

- *Fuses.* Fuses are the simplest of the switches. A fuse is a low-resistance component that provides overcurrent protection of either the load or source circuit. Its essential component is a metal wire or strip that melts when too much current flows, thereby interrupting the circuit in which it is connected. To reset the fuse, a new fuse wire of correct rating needs to be replaced.

- *Disconnects.* A disconnect is a safety on/off electrical switch mounted on the grid. While it is still a switch, a key differentiator is in the ability to physically see the separation, thereby confirming that power cannot flow through the circuit. Disconnects are required by electrical code and must be opened so that maintenance crew can see the physical separation around a device (e.g., a transformer) before performing maintenance on it. This is generally a requirement from a safety perspective. While most disconnects are operated manually, there are several places where motorized disconnects are also used.
- *Circuit breakers.* A circuit breaker is an automatically operated electrical switch designed to protect an electrical circuit from damage caused by overload or short circuit. Its basic function is to detect a fault condition, and by interrupting continuity, to immediately discontinue electrical flow. Unlike a fuse, which operates once and then has to be replaced, a circuit breaker can be reset (either manually or automatically) to resume normal operation. Circuit breakers are made in varying sizes, from small devices that protect an individual household appliance up to large switchgear designed to protect high-voltage circuits feeding an entire city.

  These are fairly sophisticated and complex devices because when current is interrupted, an arc is generated. This arc must be contained, cooled, and extinguished in a controlled way, so that the gap between the contacts can again withstand the voltage in the circuit. Given the amount of energy that must be quenched in an incredibly short period of time, manufacturers use vacuum, air, insulating gas, or oil as the medium in which the arc forms and the energy is dissipated.
- *Reclosers.* Reclosers are circuit breakers equipped with a special mechanism designed to assist with temporary faults [2]. Utilities have found that most faults in a grid are momentary in nature, meaning that the fault may get cleared automatically due to natural causes. Typical causes of temporary faults could be a tree branch swinging and causing two of the phases of a distribution circuit to touch each other, or by itself acting as a conduit to the ground. When this happens, the recloser opens up to clear the fault and tests the circuit multiple times to confirm whether the circuit is still faulted or not. Anyone who has watched the light bulbs flicker on and off multiple times during a storm has experienced the reclosed action. If the lights stay steady after one of these episodes, then it was a temporary fault, and if they stay out, then that same fault was a permanent one.

3.1.4 Relays and Protection Equipment

Two main points identify the need for relays and protection in an electrical grid: (1) the speed of electricity as it flows though the circuits, which is mainly at the speed of light, and (2) the high cost of electrical equipment and the longer lead times sometimes necessary for their replacement.

These two factors necessitate the need for extremely advanced and sophisticated set of protective equipment in an electrical grid. While electrical protection has changed a lot over the years, it goes by the general name of relays. These devices control the tripping of a circuit breaker to isolate specific parts of the grid to save the grid from being further impacted either due to a fault or other potentially major issues. There are several types of relays [3] in the network and a generic subset is listed below:

- Under/over voltage relays;
- Instantaneous overcurrent;
- Inverse Time overcurrent;
- Under/over frequency;
- Current differential;
- Impedance relays.

There are key points necessary when characterizing how these relays are set to perform their job of protecting the system [4]:

- They are generally set for automatic operation with little to no manual intervention and for constant monitoring of equipment. Many are also collecting data about the equipment that they are monitoring for post-event analysis.
- It is not uncommon to have multiple sets of relays monitoring the same equipment. This is generally done for very important/critical equipment.
- It is also not uncommon for multiple relays to have overlapping areas of monitoring and protection. This allows situations where one relay monitors one portion of the system and another relay monitors a larger portion of the system of which the first one is a part. This is often called primary, backup, and tertiary protection.
- Relays have extremely sophisticated settings that are coordinated among other relays in the network to ensure that their primary/backup roles are appropriately managed.

3.1.5 Kilovolt Classes or Common Voltage Levels

Most electrical equipment is classified by their voltage classification. This means that they are rated for a specific voltage level. For example, the class of equipment designed for use within a typical U.S. home is 110V class. In other countries, similar equipment designed for the home may be of a 220V class.

Liewise, when we see equipment on the grid, it is designated by the kilovolt class of the system where it was installed. The following classes exist but there may be more:

- Transmission equipment in the United States is generally in the following voltage levels: 765 , 500 , 345 , 230, 161, 138, and 115 kV. Some of the newer levels in transmission are in the 1100 kV class.
- Distribution equipment in the United States is generally in the following voltage levels: 69, 34.5, 13.8, and 12.47 kV.

These classifications allow the equipment to be designated on how they can be used and the level of insulation needed for them.

3.1.6 Busbars

Busbars are essential pieces of equipment used in transmission and distribution substations. These are basically either a solid strip of copper or aluminum that is used to connect different pieces of equipment together to ensure a good conductive connection between them. For example, when a transformer is connected to a transmission line, the connection is made through a busbar. Busbars can also be made of flexible stranded conductors.

3.1.7 Substations

Substations are where much of an electrical system comes together. This is the place where most of the key components of a power system are housed. In a transmission substation, we can find key components like transformers, busbars, circuit breakers, disconnects, protective relays, and so forth. Much of transmission equipment exists within a transmission substation. Key equipment outside of a transmission substation are the transmission lines, which mainly go from one substation to another. At the transmission level, substations are generally classified as based on the kind of action they perform. For example, generation substations are designed to take the power output of a generating station and step it up to transmission-level voltages. Electric power sometimes flows through many substations as it moves from supply to consumption—and through each of the substations, the voltages are either stepped up or down.

Distribution substations are not that different in terms of their capabilities. However, given that the distribution voltages are somewhat lower, these substations are also somewhat smaller.

It is fairly common for most substations to be unattended and relying mainly on some level of remote control and monitoring.

3.2 Key Power System Physical Concepts Explained

The basic electrical components covered in the previous section are very often subject to a set of physical concepts that governed their behavior. This section defines those physical concepts.

3.2.1 The Basics: Voltage and Current

The voltage [3] between two ends of a path is the total energy required to move a small electric charge along that path, divided by the magnitude of the charge. Historically this quantity has also been called "tension" and "pressure."

Voltage (measured in volts) is defined so that negatively charged objects are pulled towards higher voltages, while positively charged objects are pulled towards lower voltages. Therefore, the conventional current in a wire or resistor always flows from higher voltage to lower voltage. Current can flow from lower voltage to higher voltage, but only when a source of energy is present to "push" it against the opposing electric field. For example, chemical reactions inside a battery provide the energy needed for current to flow from the negative to the positive terminal.

Electric current is a flow of electric charge through a medium. This charge is typically carried by moving electrons in a conductor such as a wire. Current is measured in amperes and is the rate of flow of electric charge flowing through a conductor.

3.2.2 Ohm's Law

Ohm's law [4] states that the current through a conductor between two points is directly proportional to the potential difference across the two points. Introducing the constant of proportionality, the resistance, one arrives at the usual mathematical equation that describes this relationship.

$$I = V/R$$

where I is the current through the conductor in units of amperes, V is the potential difference measured across the conductor in units of volts, and R is the resistance of the conductor in units of ohms. More specifically, Ohm's law states that the R in this relation is constant, independent of the current (see Figure 3.2).

The law was named after the German physicist Georg Ohm, who in a treatise published in 1827 described measurements of applied voltage and current through simple electrical circuits containing various lengths of wire. He presented a slightly more complex equation than the one above to explain his experimental results. The above equation is the modern form of Ohm's law.

Following Ohm's law, we can calculate the losses in an electrical circuit by using the following equation:

$$\begin{aligned} \text{Power Loss } &= V * I \\ &= I^2 * R \end{aligned}$$

This means that power loss is proportional to the square of the current flowing in the circuit.

3.2.3 Kirchhoff's Law

Using Ohm's law, the next step towards calculating power flow in a circuit is through Kirchhoff's laws [5], of which there are two: Kirchhoff's current law and Kirchhoff's voltage law

1. *Kirchhoff's current law (KCL):* This law starts off from the principle of conservation of electric charge, which implies that at any junction in an electrical circuit, the sum of currents flowing into that node is equal to the sum of currents flowing out of that node (Figure 3.3).

   The algebraic sum of currents in a network of conductors meeting at a point is zero. Recalling that current is a signed (positive or negative) quantity reflecting direction towards or away from a junction; this principle can be stated as:

Figure 3.2 How voltage, current and resistance relate to each other in a typical electrical circuit.

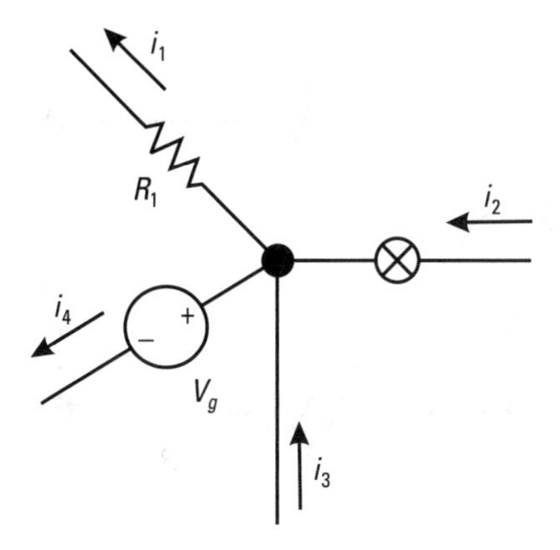

The current entering any junction is equal to the current leaving that junction
$i_1 + i_4 = i_2 + i_3$

Figure 3.3 Illustration of Kirchhoff's current law.

$$\sum_{K=1}^{n} \cdot I_{k=0}$$

2. *Kirchhoff's voltage law (KVL):* This law is based on the conservation of energy whereby voltage is defined as the energy per unit charge. The total amount of energy gained per unit charge must equal the amount of energy lost per unit charge. The conservation of energy states that energy cannot be created or destroyed; it can only be transformed from one form to another. The directed sum of the electrical potential differences (voltage) around any closed network is zero. Put more simply, the sum of the emfs in any closed loop is equivalent to the sum of the potential drops in that loop, or the algebraic sum of the products of the resistances of the conductors and the currents in them in a closed loop is equal to the total emf available in that loop (Figure 3.4).

3.2.4 DC Versus AC

The simpler definition of DC power (example of DC power is that coming out of a battery) is that is stays constant at the same level as it comes out its source of supply. AC power on the other hand is (as the name implies) alternates between a positive number and an equal and opposite negative number. The inverse of the amount of time taken by the AC wave to complete one complete cycle is also called the frequency. The frequency of the power supply in the United

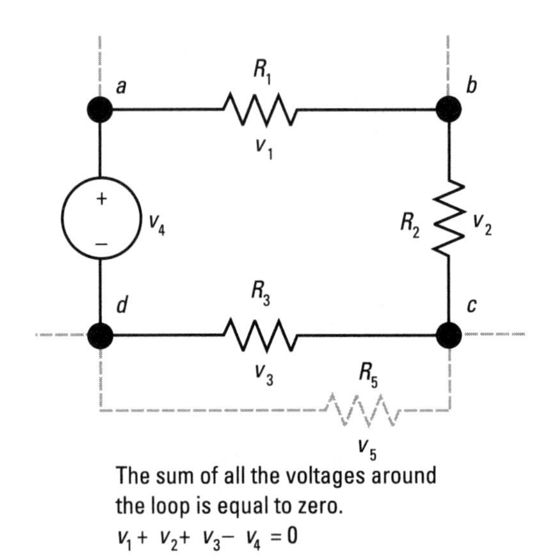

Figure 3.4 Illustration of Kirchhoff's voltage law.

States is 60 Hz, which means that the AC wave in the United States completes 60 cycles per second (Figure 3.5).

In the United States, utilities have tried both types DC and AC—even though the beginnings were actually based in DC [6], promoted by Edison.

The main difference between AC and DC leading to why AC pretty much took over from DC as the world's electricity system was in the ability to step up and step down voltages. AC could do it and DC could not—this was through the use of transformers. Some of this is now changing through the use of new technologies aided by power electronics.

Now, why is stepping voltages up and down so important? It is all about losses!

For us to understand this, we need to keep in mind two key equations that were presented in the section on Ohm's law.

1. Electricity losses are directly proportional the square of the current flowing in the circuit.
2. Voltage in a circuit is inversely proportion to the current flowing in it.

Keeping these two equations in mind, one can see that to transmit power across large distance, we should step up the voltage as high as possible to decrease the current as much as possible, leading to reduced losses in the circuit. Transformers allow this to happen in an AC circuit but there is no equivalent to a transformer in a DC circuit.

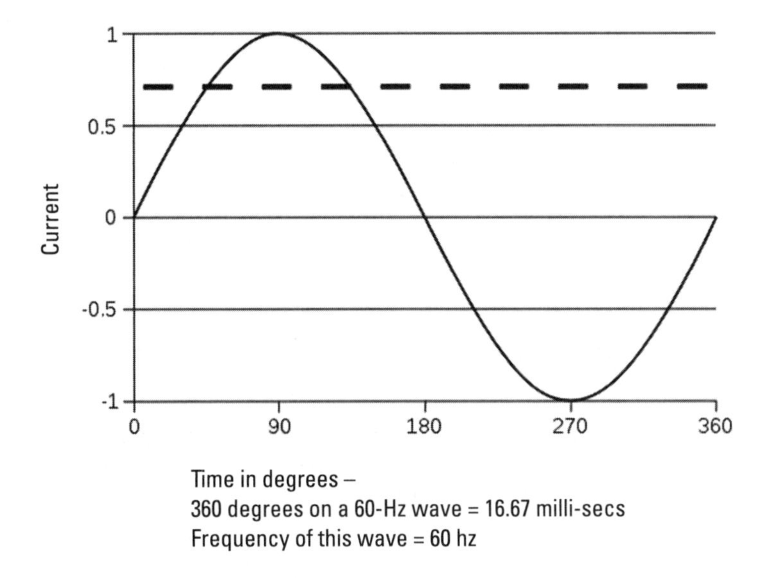

Figure 3.5 Picture of a sinusoidal wave—AC voltage.

3.2.5 Complex Power Representation

Power in an electric circuit is the rate of flow of energy past a given point of the circuit. In alternating current circuits, energy storage elements such as inductance and capacitance may result in periodic modifications of the direction of energy flow. The portion of power that averaged over a complete cycle of the AC waveform that results in net transfer of energy in one direction is known as real power. The portion of power due to stored energy, which returns to the source in each cycle, is known as reactive power.

In a simple alternating current (AC) circuit, both the current and voltage are sinusoidal. If the load is purely resistive, the two quantities reverse their polarity at the same time. At every instant the product of voltage and current is positive, indicating that the direction of energy flow does not reverse. In this case, only real power is transferred (Figure 3.6).

If the loads are purely reactive, then the voltage and current are 90 degrees out of phase. For half of each cycle, the product of voltage and current is positive, but on the other half of the cycle, the product is negative, indicating that on average, exactly as much energy flows toward the load as flows back. There is no net energy flow over one cycle. In this case, only reactive energy flows—there is no net transfer of energy to the load.

Practical loads have resistance, inductance, and capacitance, so both real and reactive power will flow to real loads. Power engineers measure apparent power as the magnitude of the vector sum of real and reactive power. Apparent power is the product of the root-mean-square of voltage and current. Engineers

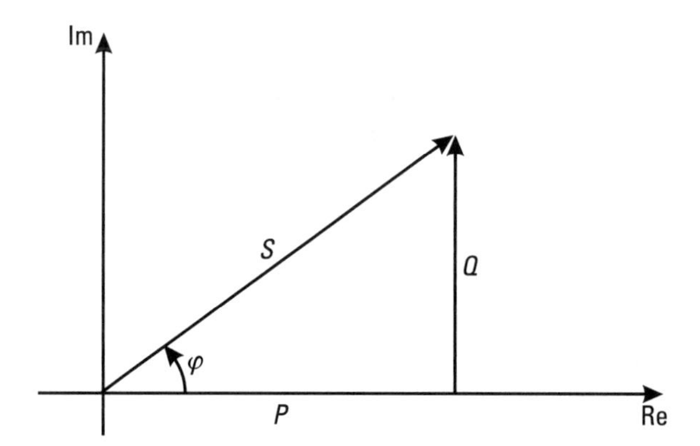

Figure 3.6 The complex power is the vector sum of real and reactive power. The apparent power is the magnitude of the complex power. P = real power; Q = reactive power ; S = complex power; |S| = apparent power; ϕ = phase of current.

care about apparent power, because even though the current associated with reactive power does no work at the load, it heats the wires, wasting energy. Conductors, transformers, and generators must be sized to carry the total current, not just the current that does useful work.

3.2.6 Power Factor

The power factor is an indicator of amount of inductive or capacitive load in the system. The best power factor value in a feeder is 1.0 and this is the value that most system planners strive to achieve. A power factor of 1.0 means that the load is completely resistive in nature or that the inductive/capacitive component of the load is neutralized.

Generally, given that the average load is more inductive than capacitive, the typical power factor of most feeders can vary between 0.8 and 0.

Bad power factor impacts the grid in several ways but mainly it impacts grid capacity because the combination of real and reactive power flows through the transmission/distribution lines, transformers, and so on while only real power is paid for (for example, in residential loads). In some commericial and industrial tariffs there are penalties for bad power factors. Improving the power factor allows more real power to flow through the various components thereby improving their utilization factors.

3.2.7 Three-Phase Versus Single Phase

The early methods of AC power transmission and distribution were all in single phase. This meant that power mainly flowed in one wire and the other wire (somewhat similar to a DC circuit) was the return or ground wire. Nowadays most power transmission and distribution is three-phase. This is very common in all the power grids around the world.

A three-phase system has three electrical conductors carrying AC currents of the same frequency but which reach their instantaneous peak values at different times. The current flowing in each of the conductors is delayed in time by one-third of one cycle of the electric current from the next one. This delay between phases makes it possible to produce rotating magnetic fields in electric motors.

Three-phase power systems have several advantages over single-phase power systems:

- Power transfer is balanced, leading to generators and motors that tend to run with a lesser amount of vibration;
- Three-phase power systems can actually transfer more power than three circuits of single-phase power and use fewer wires;
- They produce a rotating magnetic field that rotates in the direction of the phase angle differences, thereby simplifying the design of electric motors.

It is important to note that while most transmission and distribution systems are three-phase in nature, residential load is still mostly single-phase in almost all countries.

3.2.8 Phasors

Another very interesting aspect of AC three-phase systems is a physical concept called phasors. To truly understand phasors, let us consider a typical three-phase system. As explained earlier, this consists of three AC sinusoidal waveforms out of synch with each other by one-third of their cycle. This means that the waves are actually rotating around a common axis at a speed governed by the frequency of the power system that can be either 50 or 60 Hz in most countries. With a three-phase system three waveforms are rotating at the same speed but separated by a phase angle of 120 degrees.

Phasor angles are an important aspect of the power delivery mechanism because of two key points:

1. While the phase angle difference between the three phases of voltage or current at one point will be generally 120 degrees in a balanced system, they will not be in phase across all nodes in a system. This means that phase A at one end of the transmission will be different from the phase angle of the same phase A at the other end of the same transmis-

sion line. This difference is generally proportional to the amount of real power that is transmitted between the lines.

2. Just as the voltage phase angle is an indicator of the amount of real power transferred in a line, the difference in voltage magnitude is an indicator of the amount of reactive power on the same line.

These two values—voltage magnitude and phase angle—are so fundamental to a power system that their combined values are also called the state of the system. It is known that once you know the voltage magnitude and the phase angle at every node, you can calculate just about most other variables in the power system [9].

3.2.9 Superconductivity in Transmission Lines and Transformers

Superconductivity is the ability to conduct electricity without any resistance. This is a significant breakthrough in the field of electric power systems. However, until recently one needed to maintain these materials at extremely low temperatures (close to zero degrees Kelvin). More recent advances in superconductors have led to the development of or high-temperature superconductors (HTSs). These allow the ability of materials to exhibit superconducting properties at temperatures as high as 110 degrees Kelvin. As a result, utilities are experimenting with using HTS materials in transmission lines, transformers, and even distribution lines.

3.3 Key Business Concepts Explained

Superimposing over the basic electrical components and the physical concepts there are certain business concepts that drive how electric systems operate. While the examples used in this section may be U.S.-centric, these same concepts are applied all around the world.

3.3.1 Utility Interconnected System

An interconnection [10] is basically a power grid that functions at a regional scale and operates at a single synchronized frequency and is electrically tied together during normal system conditions. All of the interconnects in North America are synchronized at 60 Hz, while those of Europe run at 50 Hz. Interconnections can also be tied to each other via high-voltage direct current power transmission lines (DC ties), or with variable frequency transformers (VFTs), which permit a controlled flow of energy while also functionally isolating the independent AC frequencies of each side.

The benefits of interconnected utility systems include pooling of generation, resulting in lower generation costs; pooling of load, resulting in significant equalizing effects; and common provisioning of reserves, resulting in cheaper primary and secondary reserve power costs.

The North American Power Grid is divided into four AC electrical interconnections (Figure 3.11):

1. The western interconnection covers much of the western part of North America;
2. The eastern interconnection covers much of the other half of North America;
3. The Quebec interconnection covers the Quebec province in Canada;
4. The Texas interconnection covers much of Texas.

3.3.2 Control Area or Balancing Authority Areas

By NERC definition, a control area is an electrical system bounded by interconnection (tie-line) metering and telemetry. It controls its generation directly to maintain its interchange schedule with other control areas and contributes to frequency regulation of the interconnection.

For each of the interconnections to operate safely and reliably and provide dependable electric service to its customers, it must be continuously monitored and controlled. This monitoring and control function is distributed among the control areas that comprise the interconnection.

A control area is somewhat analogous to a pond: Water flowing into the pond is analogous to generating units generating energy, and water flowing out of a pond is analogous to energy being consumed.

In a simplified control area (not interconnected to anything else), the level of water in the pond must stay the same at all times. This is analogous to the frequency of the power system, which must stay constant. This does not change much with an interconnected power system (somewhat similar to the various interconnected systems in North America) with one exception—power (analogous to water) can flow from one pond to the other, thereby helping the levels of water in the various ponds (analogous to the frequency) in the interconnection to stay constant.

Both the generation and load in a control area are managed to keep frequency as close to constant as possible. The variation in this closed-loop control mechanism is a specific metric called area control error (ACE). This is also one of the key metrics by which the performance of the transmission/ control area operator is evaluated. Keeping ACE within the right set of limits indicates that the control area operator is doing a good job of controlling their resources instead of "leaning" on other control areas in the interconnection.

3.3.3 Renewable Energy Zones

Renewable energy zones are a new concept being considered but not yet adopted, more so in the United States. Consider the more recent proliferation of large transmission-level wind farms in some areas of the United States. Wind farms are known for their notorious volatility in power generation, which can cause havoc with control area metrics like ACE if the capacity of the wind farm(s) is a significant proportion of the overall control area generation capacity.

To combat this issue, control area operators (mainly in the western interconnection [11] are considering joining together to form renewable energy zones (REZs) [12]. The intent of these REZs is to function somewhat like a supercontrol area and share the ACE calculation across multiple control areas.

Endnotes

[1] Information on transformer and its definition are taken from wikipedia.org.

[2] Faults are caused in an electrical network when something short-circuits a live portion of the grid to the ground, causing an extremely increased amount of current to flow through the electrical network. This is considered bad for the network because it can cause components to burn out.

[3] Adapted from Wikipedia, http://en.wikipedia.org/wiki/Voltage.

[4] Taken from Wikipedia, http://en.wikipedia.org/wiki/Ohm's_law.

[5] From Wikipedia.org, http://en.wikipedia.org/wiki/Kirchhoff's_circuit_laws#Kirchhoff.27s_voltage_law_.28KVL.29.

[6] From Wikipedia: There were still 1,600 DC customers in downtown New York City as of 2005, and service was finally discontinued only on November 14, 2007.

[7] Adapted from Wikipedia, http://en.wikipedia.org/wiki/AC_power.

[8] There are a few odd frequencies in some islands that are at 25 Hz.

[9] The PMU is a device capable of measuring synchronized voltage and current phasor in a power system. Synchronicity among PMUs (or synchrophasors) is achieved by same-time sampling of voltage and current waveforms using a common synchronizing signal from a GPS.

[10] Basic information on interconnected utility systems taken from Wikipedia.

[11] These REZs are also being created in other parts of the country; for example, Texas is creating competitive renewable energy zones.

[12] http://www.westgov.org/initiatives/rtep. More information on the Western Renewable Energy Zone task force report is in this location.

4

Impact of Deregulation on System Operations

After the blackouts of 1965 and 1977, deregulation probably had one of the biggest impacts to system operations. The overall objective of system operations changed significantly to go beyond focusing on reliability on to a plethora of new tasks that were needed to support market operations. In a simplistic manner, many of these tasks were still being done prior to deregulation but mainly with utility-owned assets—transmission and generation. Now they needed to be done with several new participants all of whom needed to interface on market rules instead of a cost-based system. The expected functions to be performed by the new system operator needed to now include:

- Maintaining a reliable grid and thus ensuring the smooth flow of power from source to sink;
- Developing a level playing field for all market participants by enabling them to buy and sell power at market-based rates;
- Providing incentives for infrastructure expansion either in generation or transmission by sending appropriate price signals to the market participant;
- Providing for market-based mechanisms to purchase real-time energy through a balancing (or other type of) market;
- Ensuring adequate sources of ancillary services and compensating them at market-based rates.

4.1 A Brief Look at the History of Deregulation in the United States

The first phase of competition in the U.S. electric utility industry began with the 1978 enactment of the Public Utility Regulatory Policy Act (PURPA). This legislation required established utilities to purchase power from independent generators at prices that equaled their "avoided costs," or the cost to replace the energy if the utility needed to provide the energy with new construction. Although PURPA introduced the concept of competition into the generation of power, its passage was motivated primarily by a desire to encourage new and alternative energy-saving means of generation not by the government's intent to deregulate or restructure the industry.

In 1992, the Energy Policy Act (EPA) mandated that the industry move more rapidly toward a competitive market. It granted wholesale customers a choice of supplier while forcing the utilities to allow power transfer across their respective territories.

FERC selected Transmission Open Access (as described in Order 888/889) as the method to implement a competitive wholesale electric market. The approach chosen by FERC forced nondiscriminatory access to the transmission grid, allowing wholesale users open access to power across the grid. The objective of 888/889, according to FERC, was to provide better service and reduced rates to the end-user. As a result, utilities were required to focus on the business side of energy delivery more than ever before. The focus moved from pure reliability to adding production accountability and the responsibility for their associated financial consequences.

The FERC mandate from orders 888/889 for the electric utility industry had no historical precedent even when compared with the deregulation of the U.S. airline, gas, and telecommunications industries. The breadth of scope and the time allowed for implementation were substantially more aggressive than for the other deregulated industries. Even the word deregulation when applied to the change brought about by the FERC order only partially explained what was happening.

A more precise definition for this change could be found by examining a process with similar goals and intent that was taking place in Australia, the United Kingdom, and New Zealand around the same time. These countries had identified the process as "disaggregation"—the breaking down of the traditional vertical boundaries that once bound all entities involved in the production, transport, and delivery of power into one single organization. This breakup in Australia and New Zealand was taking a more consistent form; the separation of the entities involved in the production, transportation, and delivery of power into functionally separate units. Arguably, this better described the restructuring of the traditional electric utility industry into a more universal model. In the United States, it is fair to say that instead of creating a fully deregulated or disaggregated electric utility industry, it actually ended up creating a patchwork with full deregulation and RTOs in some parts of the country only. In other parts of the country, vertically integrated utilities still exist with functional separation built into the organization and no RTOs in others. Some locations like ERCOT went all the way into retail markets, thereby creating markets in a whole different scale there.

While Orders 888 and 889 have been credited with getting the industry moving towards electricity deregulation there have been other orders and key actions that have contributed to keeping the momentum in the right direction. Analyzing the full time of the changes in the utility industry would require one to go all the way back to the passing of PUHCA in 1935, long considered one of the more significant pieces of industrial legislation passed in the United States. Understanding the time line from then on is important to understand both the legislative calendar as well as the important events that happened along the way that lead to where the United States now finds itself (see Table 4.1). Other countries may not have gone through similar time lines or steps but over time the United States learned from the mistakes of these other countries and vice versa. From a deregulation perspective, early efforts in California both duplicated some of the successes and repeated some of the mistakes that were experienced in the U.K. market opening.

4.2 The New Participants and Their Activities

The splitting of the vertically integrated utilities happened with some anticipated set of by-products. The process resulted in the creation of a significant number of new participants from (1) other parts of the energy business (in the beginning mainly from the gas industry because they had deregulated prior to electricity), (2) restructured utilities branching out from their traditional geographic service area, and (3) the financial services industry. They brought specific resources to produce electricity better, more efficient methods for serving existing load, and new business techniques to manage risk. As the process continued, this area is still evolving especially as we find ourselves in the throes of increased potential for retail level deregulations. All participants—new and old—are continuing to become more aggressive and focused on success in this competitive market.

Figure 4.1 provides an insight into the new players who have emerged on every segment of the utility-energy value chain. Some of the new entities that would have been unthinkable prior to deregulation are players like independent transmission companies [1], and utility transcos [2].

Since the onset of deregulation, the traditional vertically integrated utility has been facing the most dramatic change. Prior to deregulation, it managed the

Table 4.1

Historical Time Line of Actions in the United States Leading to Deregulation*

Landmark Ruling/ Law	Date/ Year	Brief Description
PUHCA	1935	The Public Utility Holding Company Act is passed. The Federal Power Act is passed. The Securities and Exchange Commission is established. The Bonneville Power Administration is established.
PURPA	1978	The Public Utilities Regulatory Policies Act (PURPA) is passed, and ends utility monopoly over generation.
Energy Policy Act	1992	The National Energy Policy Act is passed.
Order No. 888	April 24, 1996	Transmission Open Access. Promoting Wholesale Competition Through Open Access Non-discriminatory Transmission Services by Public Utilities; Recovery of Stranded Costs by Public Utilities and Transmitting Utilities (Final Rule).
Order No. 889	April 24, 1996	OASIS: Open Access Same-Time Information System (formerly Real-Time Information Networks) and Standards of Conduct (Final Rule).
ISO New England begins operation	1997	ISO New England begins operation (first ISO). New England Electric sells power plants (first major plant divestiture).
California opens market and ISO	1998	California opens market and ISO. Scottish Power (UK) to buy Pacificorp, first foreign takeover of US utility. National (UK) Grid then announces purchase of New England Electric System.
Electricity marketed on Internet	1999	
Order No. 2000 (Complete Version) » Part 1 of 4 » Part 2 of 4 » Part 3 of 4 » Part 4 of 4	December 20, 1999	Regional Transmission Organization (RTO). The final rule requires all public utilities that own, operate or control interstate electric transmission to file by October 15, 2000, a proposal for a Regional Transmission Organization (RTO), or, alternatively, a description of any efforts made by the utility to participate in an RTO, the reasons for not participating and any obstacles to participation, and any plans for further work toward participation. The RTOs will be operational by December 15, 2001 (Final Rule).
The largest blackout in North American history	2003	Leaves nearly 50 million people without power in the northeastern United States and eastern Canada, some for as long as four days. A government report estimates the cost of the outage at between $4 billion and $10 billion in the United States alone.
Congress passes the Energy Policy Act of 2005	2005	The act includes repeal of PUHCA, which supporters claim will lead to greater investment in utilities. The act also creates a mandatory reliability organization with the power to impose fines on utilities. In July 2006 the government appoints the industry's existing self-regulatory body, NERC, to be the new, higher-powered reliability watchdog under the ultimate authority of the Federal Energy Regulatory Commission (FERC).
Order No. 667-A (RM05-32-001)	April 24, 2006	Repeal of the Public Utility Holding Company Act of 1935 and Enactment of the Public Utility Holding Company Act of 2005 (Final Rule).
Order No. 890 (RM05-17-000 and RM05-25-000)	February 16, 2007	Preventing Undue Discrimination and Preference in Transmission Service (Final Rule).
Order No. 729 (RM08-19-000, RM09-5-000 and RM06-16-005)	November 24, 2009	Mandatory Reliability Standards for the Calculation of Available Transfer Capability, Capacity Benefit Margins, Transmission Reliability Margins, Total Transfer Capability, and Existing Transmission Commitments and Mandatory Reliability Standards for the Bulk-Power System (Final Rule).
Order No. 745 (RM10-17-000)	March 15, 2011	Demand Response Compensation in Organized Wholesale Energy Markets (Final Rule).

*Most information has been taken from FERC Web site Major Orders & Regulations, http://ferc.gov/legal/maj-ord-reg.asp?new=sc3.

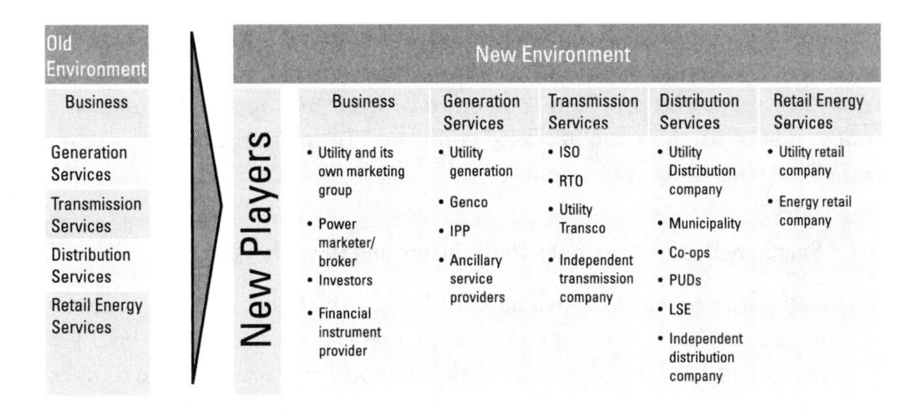

Figure 4.1 Change in utility energy chain participants due to deregulation.

entire business of marketing energy, obtained fuel for the generating plants, and generated, transmitted, and distributed energy all within a vertically integrated corporate structure. With the onset of deregulation, most utilities have needed to make extensive modifications to their structures and other companies, many of whom were already involved in the production or marketing of energy are slowly extend their business reach into this traditional arena. The changes in traditional vertically integrated utilities can be summarized as follows:

- Creation of utility holding companies to be able to better manage a combination of business units, some which are regulated by FERC, some by state PUCs, and some completely unregulated.
- Full functional separation of the generation business unit from the core utility. The utility's wholesale trading arm also moves with this business unit because they can now buy and sell in the open market where there is an RTO or not. This business for the most part is unregulated.
- Separation of the T&D business units into either one or two separate business units. The two do not have to be completely separate as would be the case with the generation business unit. This business unit(s) will need to treat the generation/trading business unit at arm's length and exactly the same as any other utility generation company, Independent power producer or other sources of nonutility generation. This unit will also still be fully regulated through FERC and state PUCs as appropriate.
- Retail and customer service still for the most part stays with the regulated T&D entities. An exception to this has occurred in states with full retail choice like Pennsylvania and Texas where independent retail companies have operated for several years.

Beyond the traditional utility, companies like independent power producers (IPPs) and other generation companies are very active in the generation arena once solely the providence of utilities. ISOs and regional transmission groups (RTGs) are now the unbiased controllers of the transmission system in areas where they have jurisdiction.

4.2.1 Summary Description of the Participants and How They Interact

The list of participants of participants in the deregulated market are many and their interactions can get quite complex. Figure 4.3 provides a high-level view of these interactions with the RTO at the center of the action. As we analyze the participants, it is important to note that not all of them will be independent companies—they could be business units within one company.

- *RTO/ISO:* A new entity with the primary responsibility of ensuring short-term and long-term reliability of grid operations. To ensure fair access to the transmission system, its management and control is completely independent of generation entities and any other market participant. Examples include Pennsylvania Jersey Maryland Power Pool (PJM), Electric Reliability Council of Texas (ERCOT) and the New York Independent System Operator (NYISO), Independent System Operator of New England (ISO-NE), California Independent System Operator (CAISO), Southwest Power Pool (SPP), and Midwest Independent System Operator (MISO). There are also several of these in Canada and in other countries.
- *Vertically integrated utility:* Like a full-service store that can offer one or more of the following functions: generation, transmission, distribution and retail, all under a holding company mechanism.
- *Generator:* Any entity that generates electric power and feeds it into the grid.
- *IPP:* An independent (nonutility affiliated) generator of energy and therefore a seller.
- *Power marketer/broker:* An entity which buys/sells energy and transmission services. With a few exceptions, this entity typically does not own generation/transmission facilities.
- *Transmission asset owner:* Owns or controls facilities used for the transmission of power. Can sometimes also be responsible for the short-term reliability of grid operations. Examples can include a vertically integrated utility, or an independent transmission company.

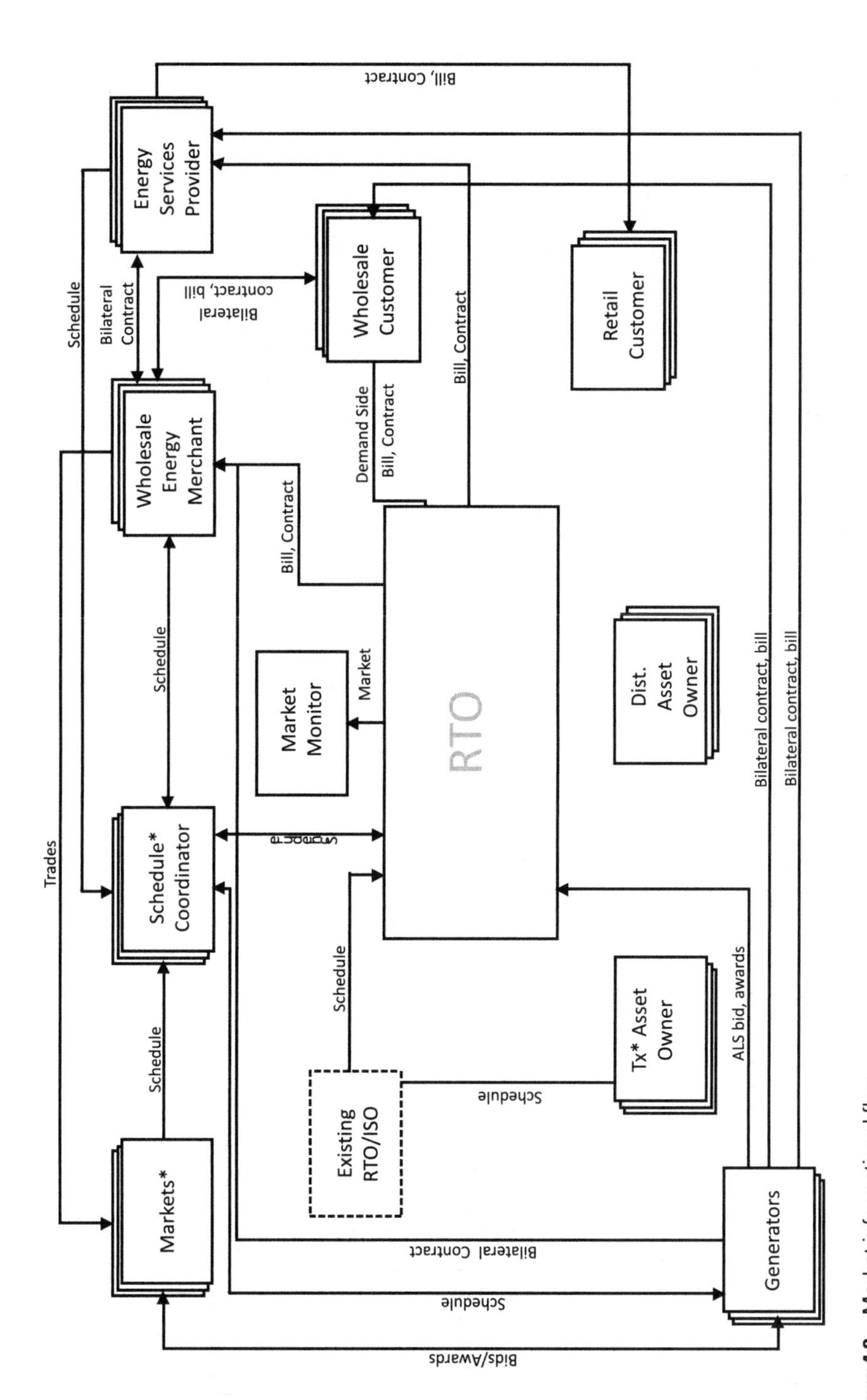

Figure 4.2 Market informational flow.

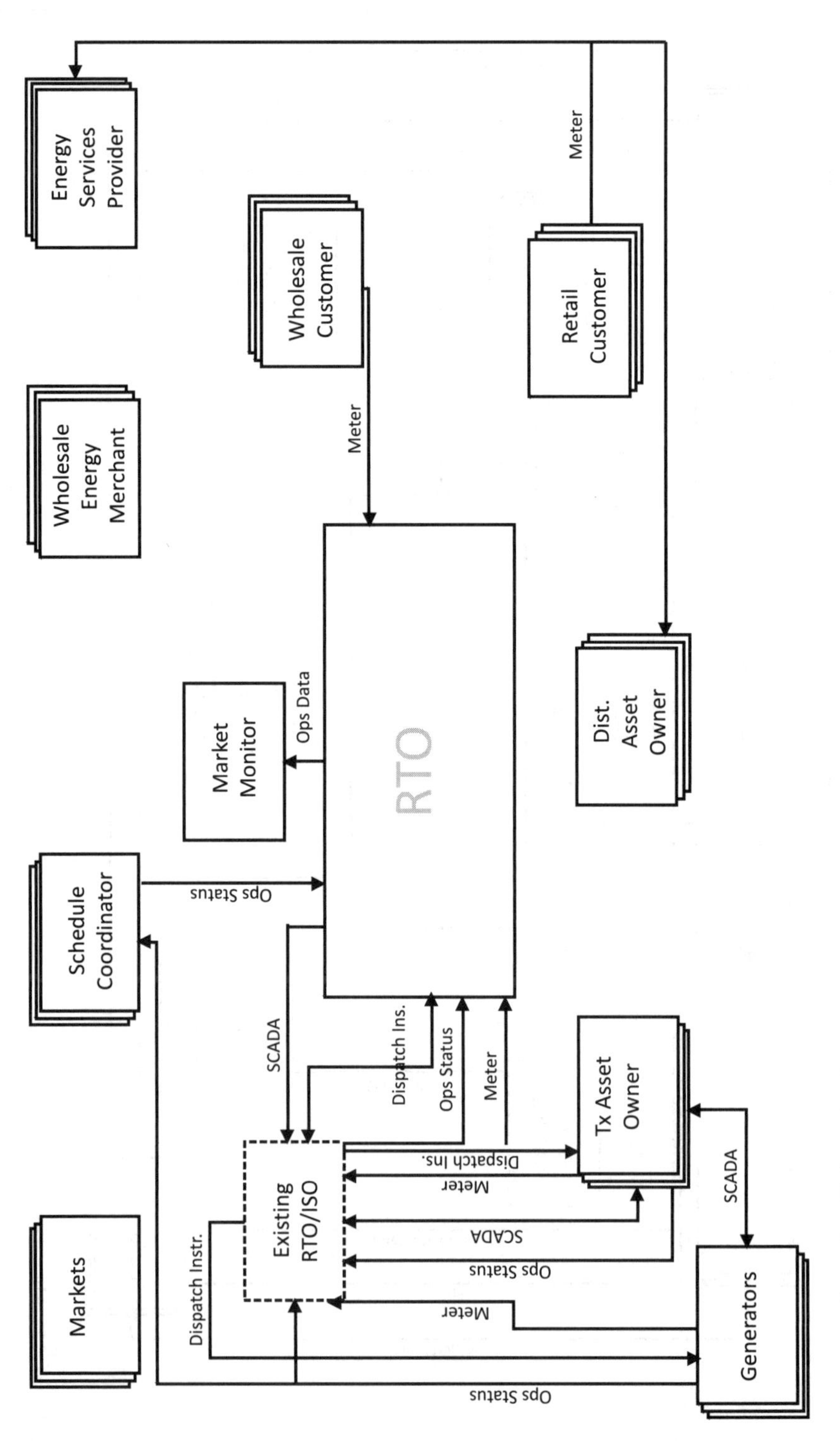

Figure 4.3 Operational information flow.

- *Transmission user (TU):* A transmission user buys the rights to transfer energy (from the seller to the buyer) through the transmission provider's transmission lines.
- *Distributed Asset Owner:* Owns, operates, and maintains distribution level substations and equipment. The interface to the small-to-medium customers that is ultimately responsible for the "obligation to serve" and may stay regulated much like the local telephone company.
- *Wholesale customer:* The role of the consumer cannot be excluded from this equation since the large consumers will (and some already are) playing a major role in buying and selling power. They are generally transmission level end use customers (e.g., large commercial or industrial load).
- *Retail customer:* Distribution level end-use customer (e.g., residential, small commercial).
- *Competitive retailer:* Retail customer interface. Responsible for arranging physical delivery and conducting commercial transactions with end-use customers but are not a part of a utility. Many terms like energy service provider (ESP), load serving entity (LSE), or retail energy provider (REP) have been used in different markets for this role.
- *Wholesale energy merchant:* Similar to ESP (above), but does not deal with retail customers.
- *Scheduling coordinator:* Responsible for creating and submitting balanced schedules and ancillary services requirements (provisions) to RTO and the OASIS [3]. Also responsible for tagging. Presently more appropriate to the ERCOT and California models.
- *Markets:* Third-party responsible for forward markets matching supply/demand, including energy and ancillary services and FTR auctions if appropriate. Can be a separate entity or run by the RTO.
- *Control area or balancing authority:* An electric system or systems, bounded by interconnection metering and telemetry, capable of controlling generation to maintain its interchange schedule with other control areas and contributing to frequency regulation of the interconnection. Where RTOs do not exist, this role is still performed by some organization who has the authority and responsibility to run a balancing market and ensure enough ancillary services are available for the system to run efficiently.

Each of these entities has a different set of roles to play in the deregulated environment.

4.3 Architectural Discussion

Figure 4.4 [4] presents the various system operations functions in an RTO grouped together in an architectural view. The capabilities are grouped and subgrouped to appropriatley identify their areas of impact. The capabilities shaded in gray are the ones that are either completely new or have changed significantly with the advent of deregulation. The others have also been impacted but to a lesser extent.

As can be seen in the figure, they can be represented in several functional blocks, all of which reside within the system operations function. These blocks are:

1. *Grid operations*: This includes the subgroups of securing the grid, performing real-time operations, and supporting real-time operations.
2. *Market operations:*Tthis includes the subgroups of facilitating the market, performing forecast, performing scheduling of energy/ancillary services, and monitoring the market.
3. *Participant operations:* This includes the subgroups of managing the participants, managing communications with them, and managing contracts with them.
4. *Commercial operations*: This includes the subgroups of gathering and managing metered data and settling the market, which also includes the billing and associated dispute resolution capabilities. The metered data identified here is not to be confused with the AMI/Smart Meter data being discussed in other places. This is generally wholesale metering aspect of the market to the extent that it impacts settling the market.
5. *Managing the assets:* Depending upon the type of RTO, this can include all the subgroups that have been listed here. In general this mainly includes the subgroup plan system enhancement. The other two subgroups build assets and maintain assets are mostly transmission owner capability subgroups and may leave some level of coordinating function at the RTO level.
6. *The two other capability groups of system administration and IT management and corporate services* are very similar to other functions at a regular utility and not as impactful of system operations and so will not be discussed in any more detail.

Let us now analyze the impacts of deregulation on system operations.

Grid operations: The main objective of this function is still network security and grid control. That aspect has not changed much from what existed under

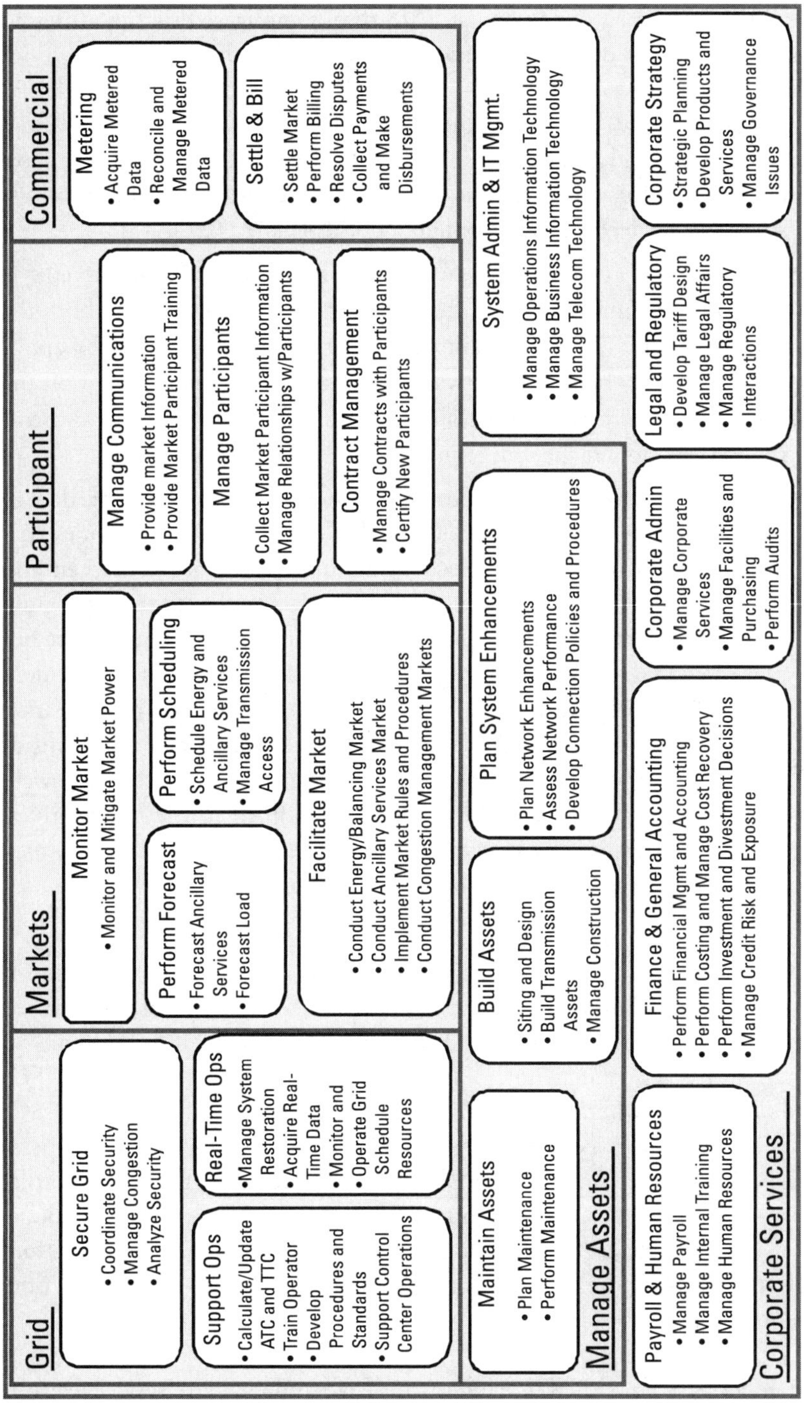

Figure 4.4 Overall system operation architecture for an RTO.

system operations prior to the start of deregulation (see Figure 4.5). These functions are still performed mainly by an EMS that is configered slightly different than that for a regular utility as opposed to a RTO. The changes are:

- *SCADA control.* RTOs are generally not allowed SCADA control of devices in the field. This action is still performed by the transmission asset owner/operator. Similarly, the SCADA data comes from the transmission asset owner/operator through a potocol called ICCP [5].
- *Calculate and update ATC and TTC.* [6] Power system conditions, system loading conditions, and the weather all have an impact on both the TTC and ATC. This calculation is performed fundamentally by applications within the EMS's network applications and then posted on the OASIS system to ensure that all market participants are made aware of the information at the same time.
- *Manage congestion.* While the system operator has always needed to solve problems of congestion, deregulation brought in new rules that governed which tools and mechanisms need to be applied when and how. Prior to deregulation, the system operator needed to run a program called unit commitment or security constrained unit commitment but as if moving the output of any generation had the same cost component to it. With deregulation, the paradigm changed towards more market forces, and in some jurisdictions they also needed to take into consideration the locational marginal aspects of the loading and flows as well. While the mechanisms and tools are well defined in the market rules, the actual levers in terms of which resources to move would be now provided to the system operator by market operations.

Market operations: This is a completely new set of capabilities that came in to the control room after deregulation was implemented and most of it was as a result of managing the market (see Figure 4.6). For a typical utility many of the functions listed here may not apply—they apply more to a RTO or an ISO. However, in places where a market (outside of jurisdictions of PJM, MISO, SPP, ERCOT, ISO-NE, NYISO, and CAISO) does not exist, the system operator is expected to run some basic market functions to run a balancing market [7]. A key part of the balancing authority is also to provide generation support where market signals are sent to all generators participating. In this section, we will only focus on those portions that are either perfomed by or impact the system operator.

- *Performing forecast.* While forecasting load and ancillary service was important even prior to deregulation, the biggest difference came from the

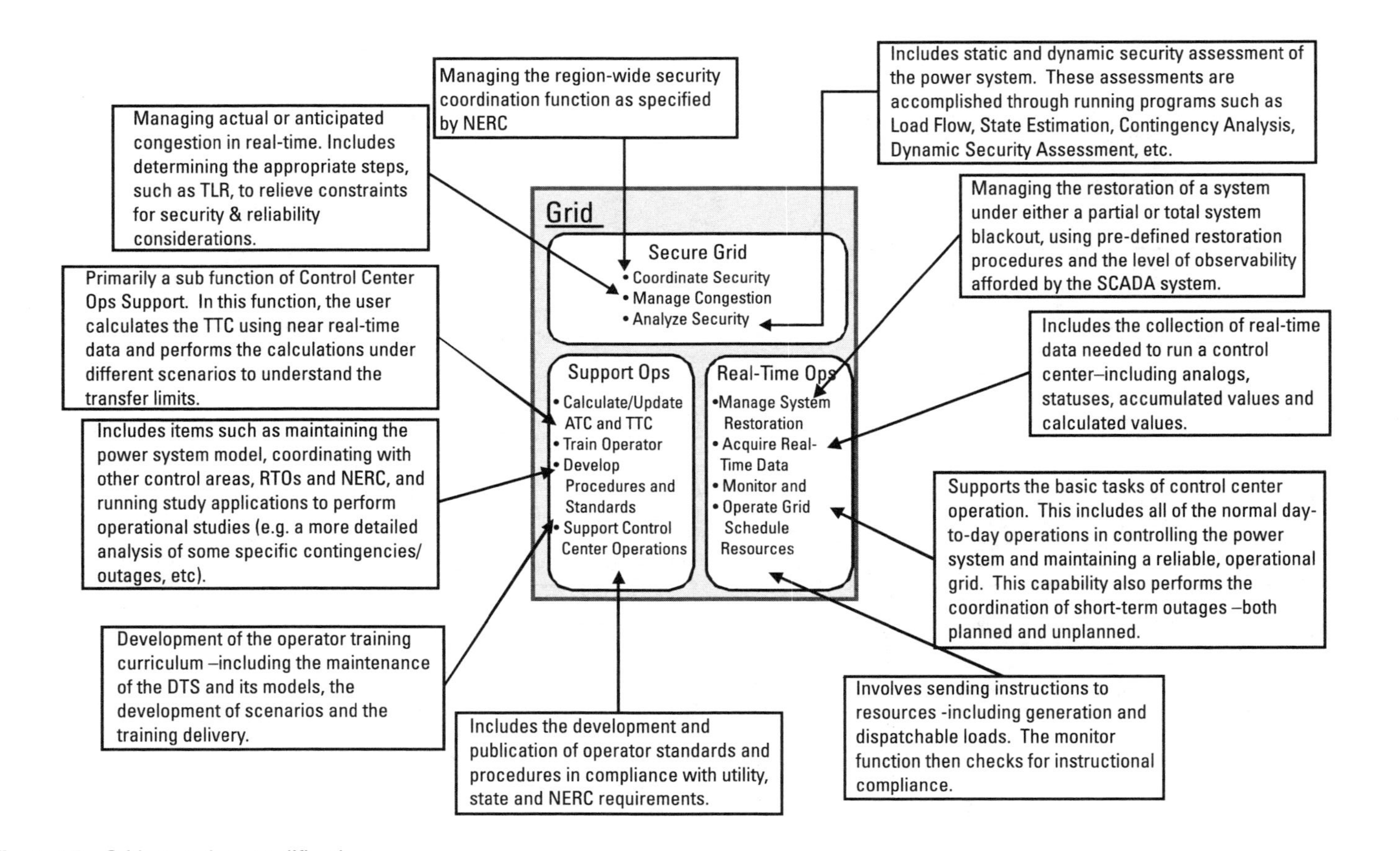

Figure 4.5 Grid operations modifications.

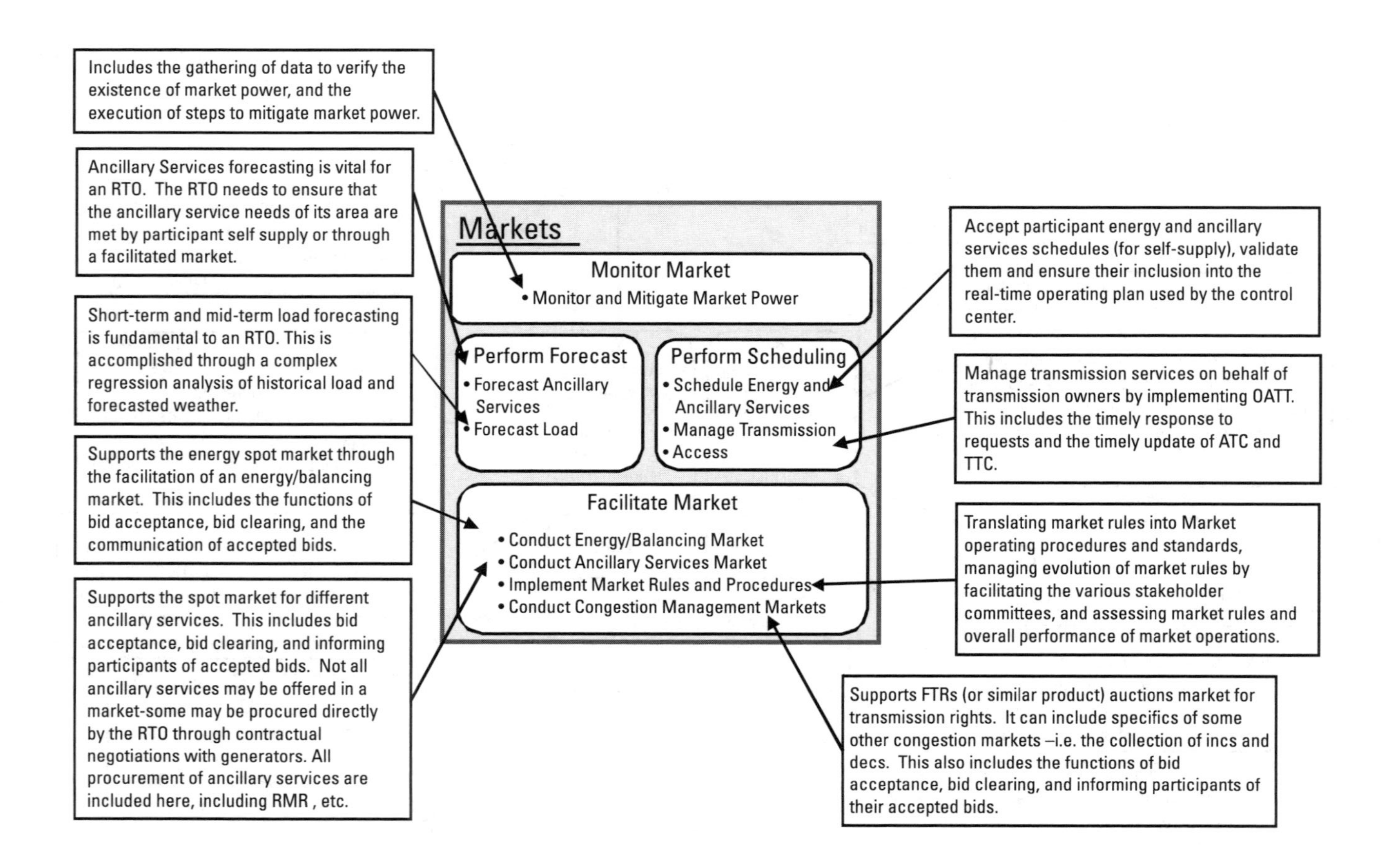

Figure 4.6 Market operations impacts to the grid operator.

fact that this forecast would not be used to set the market in terms of how much would need to be purchased in actuals and how much in reserves. All of this also impacts market participants because once the market clears for them, there is a revenue implications on the outcomes. Accurately predicting these values based on system, weather, and other conditions is important.

- *Facilitating the market.* Different jurisdictions run different markets based on diferent rules. Some states like Texas also run a retail market at the same time. Most run a locational marginal pricing-based market as well as a combination of a day-ahead, day-of, and real-time markets for energy and ancillary services. The actual inner workings of a market are not that relevant from a system opertor perspective with the exception that they all take the cost of alleviating congestion into consideraton as they clear the market.
  When there is no market in a jurisdiction such as in the Northwestern United States, much of the energy transactions are done through bileateral trades between specific parties and this entire process is facilitated through a balancing authority who is held responsible to ensure that sufficient supply has been made available to service the load. The balancing authority also operates a balancing energy market to ensure that appropriate reserves and ancillary rvices are made.
- *Performing scheduling.* The outcome from a market clearing process are a set of energy and ancillary services schedules that are fed to the system operator who then uses them to manage and operate the grid efficiently. As grid conditions change during the day, they are reevaluated in the real-time market and then readjusted.

Commercial operations: This represents another new set of functions in a system and depending on whether the operator functions as a normal transmission/distribution operator (see Figure 4.7), RTO/ISO, or a balancing authority, some of these capabilities will still need to performed either for other participants n the market or for themselves to perform a shadow settlement of the market operator to ensure that they are able to collect on what is due to them. However, these functions have no impact to the system operator or vice versa.

Participant interface: This is the last of the set of totally new capabilities that got introduced into the system operator function due to deregulation (see Figure 4.8).

Much of these capabilities will only be needed in a system operator if they are functioning as a RTO/ISO or a balancing authority. When the system is also performing those functions, the need to manage participants and contracts with them becomes an important part of the system operator.

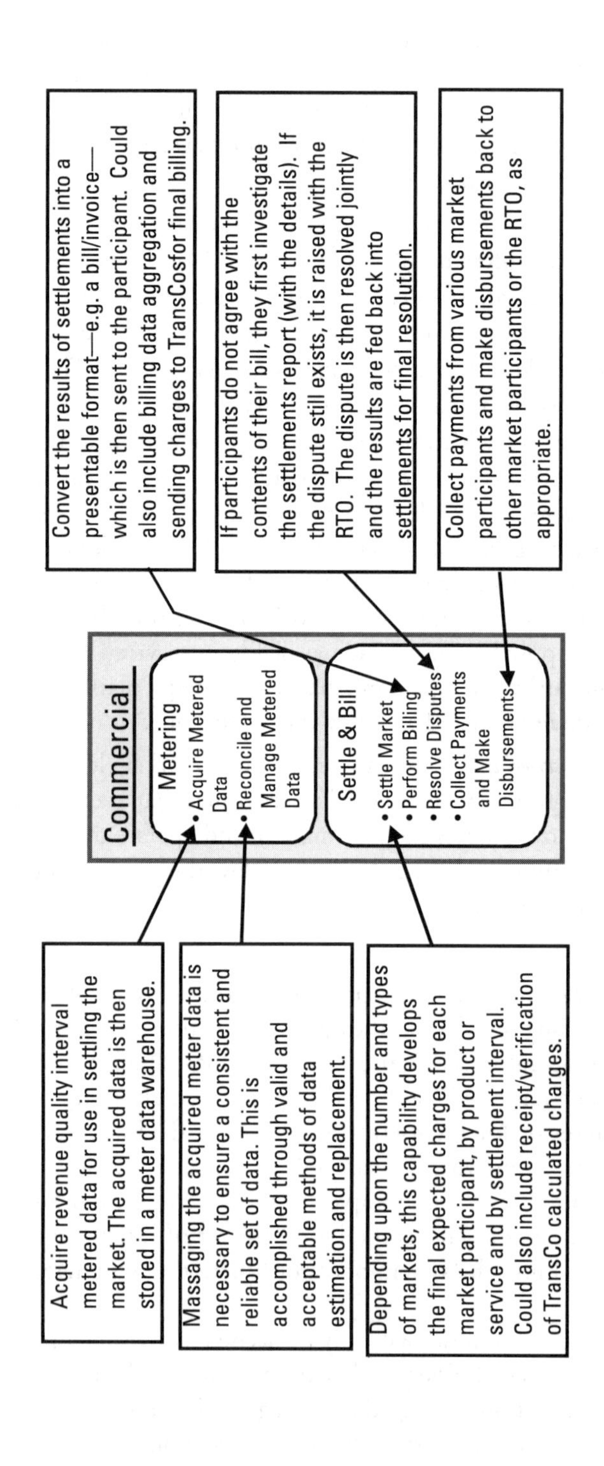

Figure 4.7 Commercial operations in a system operator.

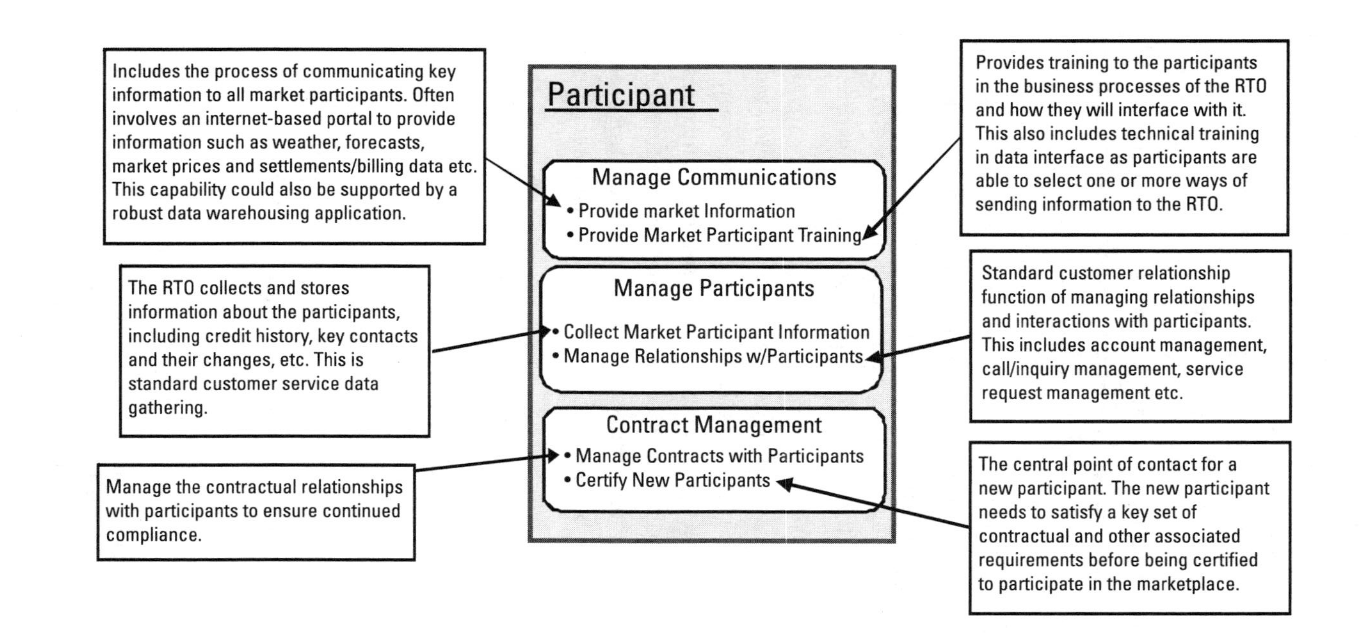

Figure 4.8 Participant interface functions in a system operator.

4.4 Conclusion

Deregulation brought along some of the most dramatic impact on system operations. As can be seen in this chapter, much of it impacted transmission system operations with the exception of places like Texas, which also went deep in to retail choice where there are some impacts to distribution operations as well.

The key changes can be summarized as:

- Brought a commercial mindset into the control center with focus on settlements and the need to perform financial transactions, which had a significant impact on their utility's bottom line.
- Brought a customer service mindset into the control center because of the need to deal with players who were not employees of the same utility, were not of a reliability mindset like themselves, and who very often had very different corporate objectives than a typical utility. Employees of these companies could sometimes even be competitors and so all the interactions needed to be strictly process driven, transparent, and auditable.
- The move towards making transmission tracking and availability transparent (OASIS) required a real Chinese wall-like separation between core control center functions and transmission functions, which also needed to be separate from generation and trading functions, required a major redesign of all of these capabilities in the control center or needed a single utility to buy more than one set of system operations technological solutions.
- As defined above, core functions like generation dispatch and unit commitment that used to be all cost-based now became market-based, leading to very different patterns in how generation was either dispatched, made available, or compensated.

As retail choice becomes more prevalent, it is possible that these changes will also contine to extend from transmission into distribution operations.

Endnotes

[1] An independent transmission company (ITC) is basically an entity that would operate as a stand-alone, for-profit transmission business. FERC is encouraging ITCs to function in this marketplace in conjunction with RTOs that meet the requirements of Order 2000.

[2] A utility transco is basically the IOU spinning off its transmission business unit into an independent company. AEP transmission performed this action more recently in 2010.

[3] The Open Access Same-Time Information System (OASIS) is an Internet-based system that was originally developed in 1996 in response to the mandate of FERC orders 888 and 889. The intent of this system (as the name implies) was to make transmission services available to all wholesale market participants at the same time. This was an important movement because for the first time, even the vertically integrated utility that owned the transmission line needed to go to the same site to reserve for their transmission service needs in the same way as a merchant generator. By reserving transmission services, anyone could use (or resell) transmission services to move their energy from source to consumer.

[4] The picture presents more of an architectural view. Depending upon whether the entity is an RTO, vertically integrated utility, a disaggregated utility (as in Texas), each version of system operations will have more or less emphasis on the various capabilities either old or new.

[5] The Inter-Control Center Communications Protocol (ICCP) is a very specific communications protocol used by utilities worldwide to exchange data over the wide area networks in a secure and real-time manner. This data protocol is now used by a combination of utility control centers, RTO/ISOs, utilities, regional control centers, and nonutility generators.

[6] TTC stands for total transmission capability, which is the total transfer capacity of the transmission corridors available for the wholesale market to take advantage of in order to move their generation supply to the locations of the load. This information is posted on the OASIS system and wholesale traders make reservations requests for their needs. ATC is the available transmission capability, which is the TTC minus the amount of capacity that has already been reserved.

[7] Electricity load fluctuates constantly, and any changes in load demand that are not offset by a change in resource schedules (in essence, under- or overscheduling) require the balancing authority to meet the demand by purchasing electricity from generators or by compensating those to reduce. In either case, the expense is recouped from the load-serving entities. This additional power (or reduction in power) is purchased in the "balancing" market. The balancing authority determines a market-clearing price every 15 minutes that it will pay to generators that sell energy in the balancing market. The market-clearing price is the price paid for the last megawatt procured in the bid-stack for balancing energy and is paid to all generators providing this service.

5

Impact of Smart/Modern Grid on System Operations

Intelligent electric networks are defined as a state of increased awareness in a network and its ability to respond in real time, leading to better operational effectiveness for the utility and an improved experience for the customer. It represents a complete transformation of today's electric grid.

The electricity system in the United States received renewed attention after the August 2003 blackout that affected more than 50 million customers across the Northeast United States and caused billions of dollars of damage to the U.S. economy. This blackout became a call to action as the event exposed the United States' dependency on a vulnerable infrastructure. This situation was found to be unacceptable and changes were needed to ensure increased safety and reliability across the system.

The smart grid is one of the results of that call to action.

The utility industry is embarking on a transition whose end isn't fully understood. That uncertainty is due at least in part to the fact many utilities and regulators aren't yet comfortable relying on smart grid technologies as a substantive resource planning tool. This isn't stopping utilities and vendors from making those investments, but companies and regulators are questioning the assumptions used in smart grid investment planning. Some analysts point out that experience teaches that the more complex a system becomes, the more fragile it becomes. Many people in this mature, bricks-and-mortar industry generally distrust modern technologies as a resource planning solution. They think we are putting too many eggs into the smart-grid basket, and we will regret it as we did our heavy reliance on nuclear and gas during their boom periods in recent decades.

However, the underlying problem has not gone away. As industrialized societies continue growing, worldwide electricity demand is estimated to double by the year 2030, and the minor inconveniences that customers currently notice in the power grid will increase, becoming more pronounced and problematic.

In addition, the notion of expanding power capacity simply by building new power generators cannot be offered as a reasonable alternative, as obtaining building permits for these new facilities is increasingly difficult. In addition, consumers and stakeholders are pressing for productivity increases to accommodate demand growth and rising capital costs. Users are expecting quality, reliability, and power production increases on the one hand, while at the same time demanding that the electric power industry reduce or mitigate its carbon emissions and increase energy efficiency. Managing the grid will become more complex with implementation of state renewable portfolio standards (RPS), which will require that utilities use more renewable sources of energy.

Figure 5.1 presents a futuristic conceptual look at the smart grid and how it could impact the generation, delivery, and consumption of energy. As is visualized in this picture, in the future we could foresee a combination of diverse energy sources both at the bulk/wholesale level as well as at the distribution level, delivering power through a smart infrastructure to a combination of smart homes and smart buildings. This future could hold a combination of energy sources such as hydrogen (fuel cells) or biofuels. We could envision grid-scale storage making a play, supported by distributed storage at the community level such as community energy storage. At the retail end smart homes and smart buildings become willing participants in the grid of the future through a combination of automated and semiautomated control of their consumption. It must be noted that much of this is in a pilot-like stage and being proven in various jurisdictions around the world. The pilots are trying to prove their effectiveness both from functionality as well as economic viewpoints. Putting the delivery system in the middle gets to the heart of this picture because the system operator of the future will need to be able to manage this diversity of supply and consumption and at the same time provide the flexibility to the consumers.

The key drivers for a smart grid (in addition the 2003 blackout), for example, are:

- *Customer expectations.* Customers are demanding higher levels of service. Utilities need to ensure that their service at least matches, and perhaps surpasses, the standards set by other industries.
- *Workforce skills shortage.* The utility workforce is aging. It is anticipated that the industry could lose half its skilled workers in the next 5 to 10 years to retirement.

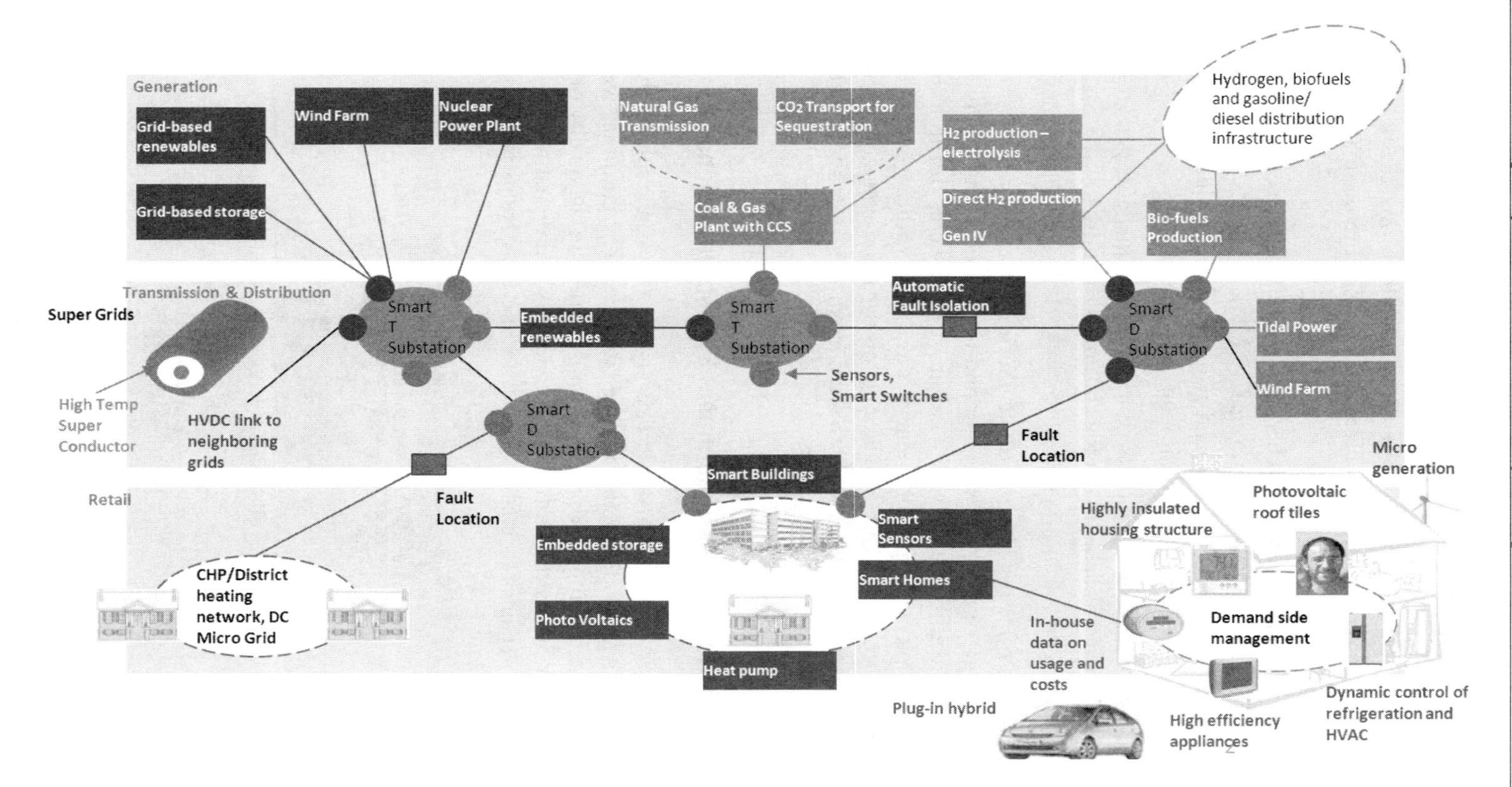

Figure 5.1 A conceptual view of the smart grid. (© Accenture. Reprinted with permission.)

- *Infrastructure replacement.* Aging physical infrastructures will require extensive replacement, and those costs are expected to outstrip historical capital spending rates by more than $14 billion over the next 10 years. There is growing recognition that utilities need to be smarter in implementing replacements.
- *Environmental constraints.* There is a concerted movement toward RPS and reductions in greenhouse gas (GHG) emissions. This fact, combined with rising fuel costs and rising load, is making the delivery of power more expensive and difficult.
- *Technology.* Technology costs continue to decline, and performance of technologies available to deploy continues to improve.

All of these factors [1] necessitate the implementation of a smart grid capable of monitoring the transmission and distribution and able to regulate transmission and distribution when the smart grid senses disruptions in the system. The implementation of a smart grid, along with an overall demand management system, will help provide the lower-carbon-emissions future that has become a necessity in terms of both our current standards and future energy requirements.

5.1 Smart Grid Changes That Impact System Operations

The key changes on the utility side (see Figure 5.2) as a result of the smart grid are basically invisible to the average customer. The customer would see the results of these changes, which is an improvement in the quality of service. On the transmission and distribution side, sensors and digital relays are being installed on power lines to enable utilities to operate systems with greater ef-

Figure 5.2 Conceptual view of distributed energy delivery to a home. (© Modern Grid Solutions.)

ficiency and reliability. Today's SCADA systems, for example, typically provide data on the state of transmission or distribution every 2 to 4 seconds. PMUs are being added to the transmission system that can sample voltage, current, and many other variables several times a second or faster—giving utilities and system operators a far more accurate view of the health of the grid. A broad deployment of synchrophasors are planned to be used in a new kind of systems called WAMS to become an early warning system to help halt or prevent power surges before they develop into massive blackouts.

Similarly in a distribution system, other sensors and controls like IVVC, reclosers are being added to provide more visibility and control of the flow of energy and the state of the system in terms of quality of service to the customer.

These digital sensors and remote controls to the transmission and distribution system would make it smarter, greener, and more efficient, resulting in what is being called as an energy Internet. The expectation is that this new system would be far more responsive, interactive, and transparent than today's grid. It would also be able to cope with some of the other new changes that are being done to the grid—integrating new sources of renewable power, support charging of electric vehicles, provide information to consumers about their usage, and allow utilities to monitor and control their networks more effectively.

There are other smart-grid technologies that would be more visible to the customer. Probably most important would be the introduction of smart meters, which track electricity use in real time and can transmit that information back to the power company. Smart meters have been used by commercial and industrial customers for decades, but in recent years are slowly becoming cheaper to support wider deployment.

5.2 Community Changes Impacting System Operations

The electric energy user community is also making several changes to their life styles that impact utility operations significantly through the need for smart grid technologies, as outlined below.

5.2.1 Proliferation of Distributed Generation, Renewable, and Community-Level Storage

New sources of localized generation sometimes supported by storage are making a slow and steady move into neighborhoods. Generation can either be home-based gasoline or natural gas gensets, or could be of the renewable kind. The most common generation of the renewable kind is solar PV cells and sometimes wind-powered. Storage can either be at the home or of the community energy storage (CES) [2] kind.

Distributed generation or storage of any kind impacts distribution system operations in several ways. First, it brings in new sources of generation that could be somewhat unpredictable, leading to fluctuations in power quality to the customer and/or backflow to the grid. It also could also represent a safety and protection issue due to its ability to cause two-way power flow in a system that is basically designed for one-way power flow.

5.2.2 Advent of Electric Transportation

Electric cars and other similar forms of transportation are making their mark in our society (see Figure 5.3). Some key examples of these vehicles showing up in residential drive-ways include the Nissan Leaf, Toyota Prius PHEV, Chevy Volt, and others. As plug-in hybrids and all-electric vehicles become more popular, they will be added to households already stuffed with power-hungry devices ranging from big-screen TVs to electric dryers. Adding a plug-in car to the grid is equal to about a third of the load of a house [3].

The key issue from a system operations perspective is that these cars represent new load that tends to bunch up in some neighborhoods and can cause overloads of distribution lines. An equally important issue is associated with the life of distribution transformers at the local level. These transformers are designed to be used heavily during the daytime and cycle down in the night when the oil is used as the insulator and coolant for the transformer. If electric cars (that require charging at night) perform their activity for about 4 to 8 hours every night, the transformers will not have a chance to cool down—potentially leading to shorter run times, more replacement costs, and more outages.

5.2.3 Introduction of Microgrids

Microgrids are slowly making their way into the modern grid, mainly showing up in industrial campuses, university campuses, and military bases. By their very definition, they are designed to either operate independently, bringing together a variety of sources of generation to service their load, or to be fully connected into the grid.

This very feature that is their strength is also their complicating factor with respect to system operations. When connected to the electric grid, all of their fluctuations need to be handled by the system operator who is still required to deliver high-quality electric supply to them. Microgrids also present the complexity of bidirectional power flow that needs to be managed for the safety and reliability of the grid [4].

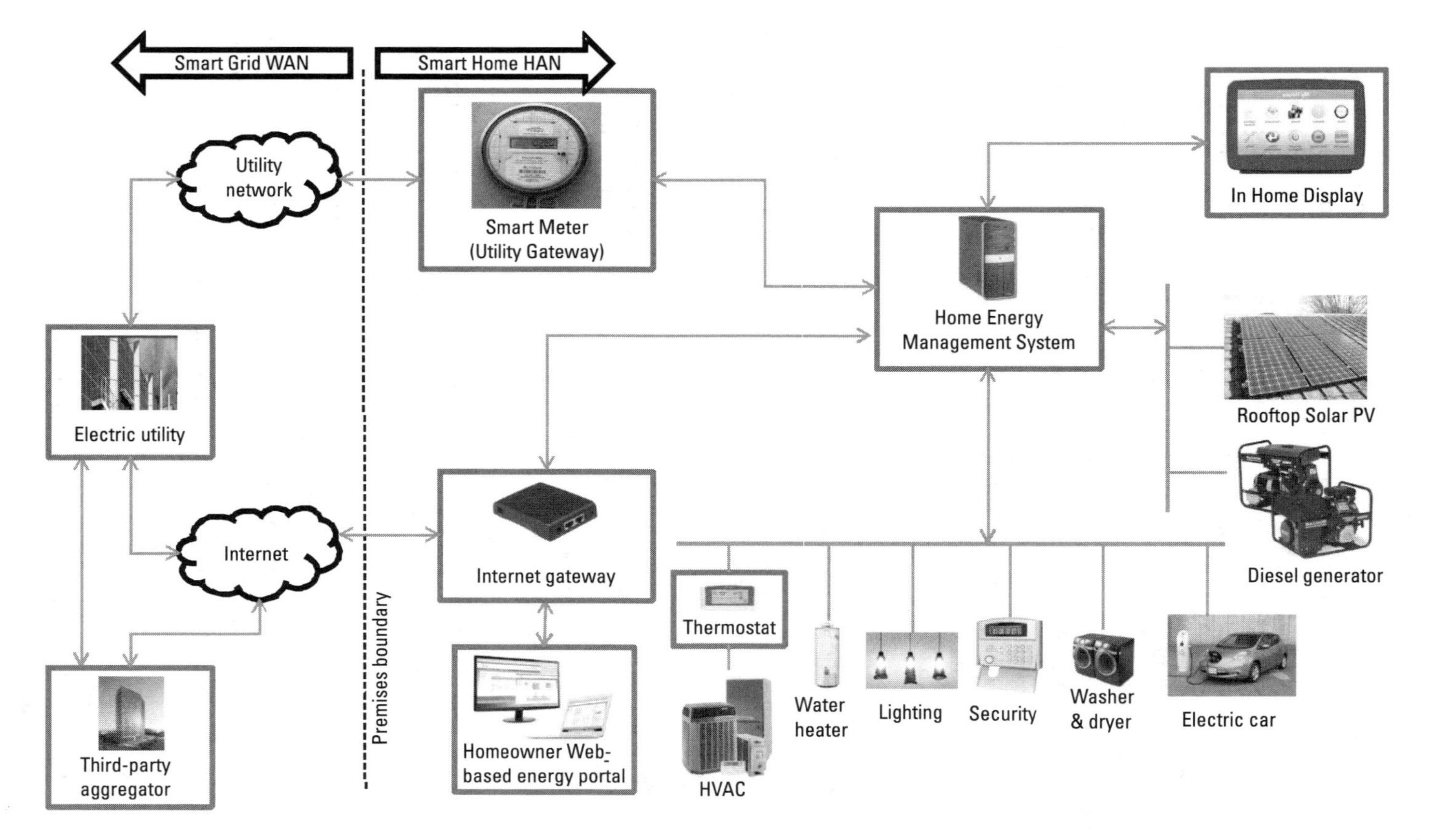

Figure 5.3 An illustrative example of a home area network. (© Modner Grid Solutions.)

5.2.4 Introduction of Smart Appliances

Smart appliances [5] are on the cusp of transitioning from the development to commercialization. Appliance manufacturers are working with utilities to test various capabilities under different scenarios. It is anticipated that smart appliances will have an acceptance curve somewhat similar to that of the "Energy Star" appliances. If that happens, then it is quite probable that percentages of smart appliances sold in the marketplace will slowly increase to become the default within the next 5 to 10 years (see Figure 5.3).

From a system operator perspective, the key here is to leverage smart appliances to significantly improve the stability and operational efficiency of the electrical grid with limited impact on the lives of energy users. The system operator's ability to manage discrete device components to reduce energy consumption at a moment in time allows him or her to offer load management option such as demand response programs. System operators need to have tools that allow them to take advantage of these mechanisms.

5.3 So What Does This All Mean to the System Operator?

For the system operator, smart grid changes everything. The magnitude of change is very similar to the types of changes that happened during deregulation. Some of the key changes are outlined below.

- Given the plethora of sensors and controls, there is expected to be a paradigm shift from a centralized command and control mechanism to an advanced network of both centralized and distributed integrated systems that can make intelligent decisions.
- At the transmission level, the advent of PMUs is bringing in new tools that provide far more information to the system operator both at the predictive and at the reactive level. It is anticipated that transmission will still stay centralized.
- Distribution system operations will move from a paper and pin version of system control from back rooms of utility service centers to more formal command and control centers using systems called DMS, OMS, SCADA, GIS, and others. It is anticipated that distribution will convert itself into a combination of centralized and distributed control mechanism.
- The system operator needs to move from dispatching a centralized set of controllable generators to a combination of centralized and locally distributed generation sources, some of which are controllable and some of which (like wind and solar) are not (Figure 5.4).

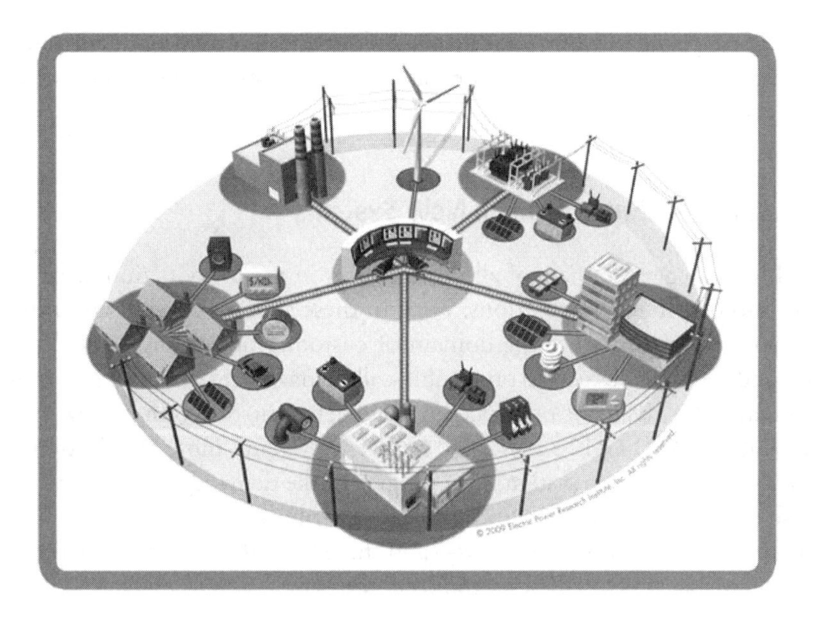

Figure 5.4 A control center view of the smart grid. (Source: EPRI.)

- Improved sensors, which can instantly observe the state of the grid and transmit the information to different locations.
- Entry of new technologies (like virtual power plants), which through the use of smart grid technologies (like AMI, Demand Response, and other controls) is able to provide virtual power and improve the efficiency of the overall system.
- More automated processes supported by trained people.
- Adding a whole new dimension of cybersecurity and privacy to operations. Given the plethora of sensors, smart meters, and similar products that are anticipated to be added on to the network (and whose data may be coming into the control center, some of which will be controlled by the system operator), the area of cybersecurity is getting a whole new look. For the first time, utilities are looking at large-scale usage of public networks for the purposes of communicating with these new devices. Cybersecurity standards are being developed and should become normal and accepted in the near future. Similar to the focus on cyber security, privacy is becoming an issue primarily due to the focus on demand response and the ability of the utility (or other entities like aggregators) to go onto the premise and control loads inside. Privacy advocates are demanding increased levels of attention on the use and widespread availability of personally identifiable information (PII) and are working on developing standards for them as well.

These changes will breathe a tremendous amount of new life into electric system operations and make the operators more capable of making better decisions faster and cheaper.

5.4 Impact of Smart Grid on New Systems

Possibly the biggest impact of all these areas is in the introduction of new systems into distribution operations. Prior to these systems coming on, distribution operations was either the domain of customer operations or field operations and the main focus was on trouble-call management, which has been long considered the precursor of outage management. Introduction of new sensors and controls, AMI and its associated head-end systems, more SCADA into the distribution end of the system, new programs like demand response and time-of-use rates have required utilities to make distribution operations into a more formal set of operational functions—and these have all been supported by new sets of systems, which will be introduced next.

5.4.1 Meter Data Management System

Meter Data Management (MDM) is a system designed to store and manage data collected from electric meters. Originally, the data was collected manually and entered, but with the advent of smart meters, this activity is now done automatically. The data consists of voltage, current, harmonics and so on. Primarily, usage data is brought in from the head-end servers that serve as the main collectors of the data from the meters.

The MDM system imports the data, validates it, checks it for errors, and then processes it before making it available for billing and analysis. As smart meters are becoming more prevalent, the data contained in an MDM system is used for far more than customer billing; it is also used for outage management, asset management, and analysis of utility operations.

With anticipated increase in the number of meters in service supported by better communications, it can be expected that these meters could end up sending their data to the head-end system more often. As these changes occur, system operators are anticipating that this could be an opportunity for meter data to be routed to the DMS as well to provide better and more complete ability to observe the system in more detail.

5.4.2 Outage Management System

A significant portion of the North America's distribution networks are still not telemetered. As a result, utilities depend on customers to report outages; customers need to call in and report their outages. An OMS collects, analyzes calls, and determines probable device failures and/or probable outage locations. Using this information, utilities are able to reduce the time to identify, prioritize and resolve network incidents, and effectively communicate with their customers and regulators.

From a value perspective, OMSs allow the utility to reduce the duration of outages (as measured by SAIDI and CAIDI), improve customer communication before, during, and after outages, and improve operational efficiency by reducing overtime, dispatch centers, and storm staffing.

OMSs had their humble beginnings in trouble-call management systems (TCMSs), which were created to process the inbound avalanche of calls that follow a power outage and to manage the outbound communication of that same information to field crews, customers, stakeholders, and others who require it. As more and more telemetry was available from systems like SCADA, trouble-call systems became more complex and developed into OMSs.

In several implementations, MDM systems are being integrated with OMS systems so that outages can be more automatically identified instead of solely relying on customers to call and report their outages.

5.4.3 Geographic Information System

GISs are designed to capture, store, manipulate, analyze, manage, and present all types of geographically referenced data. In the simplest terms, GIS is the merging of cartography, statistical analysis, and database technology.

Fundamental to a GIS is in its ability to store and correlate disparate types of data like type of device, ID, location (GPS coordinates), characteristics, and connectivity with their geospatial rendering. While this type of information was not that important when trying to model and support transmission systems (due to fewer components), it is tremendously important to support distribution systems.

From a system operations perspective, the GIS is a fundamental system that is designed (among other things) to deliver as-built static and physical data to OMSs, DMSs, and other similar systems that depend on an accurate representation of various power system components in the field and their connectivity.

5.4.4 Distribution Management System

A DMS comprises a base SCADA system that is equipped with additional planning and operations functions for the utility's subtransmission and distribution feeder systems. DMS applications are highly data-intensive. This is due to the greater number of power system elements and spatial information to be included in displays, analysis functions, and databases.

The DMS enables distribution system operators to manage their responsibilities of monitoring and operating the grid, coordinate clearances, create switching orders, and support emergency and storm management. As a system, the DMS for the most part is the distribution equivalent of the EMS, which was developed for transmission. The advent of smart grid has made the DMS the cornerstone of the system operator's response for everything that the smart grid can throw at the utility.

5.4.5 Distributed Energy Management System

As DEMSs are getting more and more accepted in the industry, an important gap has been identified by most utilities and vendors. This is the gap between the distribution tap-end transformer and the customer: With so much going on at the customer end of the system under the auspices of the smart grid, new capabilities are needed at this end. For lack of an industry-accepted name for this system, we have called it a distributed energy management system or DEMS. A typical DEMS system does not need to perform the full suite of power systems applications but has a combination of SCADA interfaces supported by one or more optimization mechanisms to support specific applications. We also believe that this system could also be a path for those (generally) smaller utilities that either do not need the sophistication of a full-fledged DMS or cannot afford one.

Two major application suites stand out from a DEMS perspective. It is important to note that it could either reside independently, be a part of the DMS, or be integrated into the DMS.

5.4.5.1 Demand Response Management

There is a need to track various demand response (DR) programs and the customers who have signed up for them. The system would enable communication of the various DR triggers to the participating customers and track their responses, all in one place. This system will perform all actions required by the DR programs. The data collection and the associated analytics tend to be an added bonus in these systems with a possible outcome being reports that can track compliance to the program requirements and their effectiveness.

5.4.5.2 Distributed Renewables Management

As energy storage mechanisms, electric vehicles charging, renewables, and other forms of distributed generation proliferate, distribution operators see a need to track and manage them. Key to this application suite is the ability to track their generation/consumption, the different controls that can be sent to them, and the forecasting that would assist the system operator to manage the loading on the system appropriately.

Endnotes

[1] Implementing a "smart grid" is becoming an increasingly common topic. At both the state/federal level and internationally, initiatives are under way to modernize the electric transmission and distribution system, and the discussion has extended beyond the traditional industry channels into the mainstream media. For many, the idea of a power system that can automatically provide for the needs of all stakeholders at all times is a compelling vision.

[2] [http://www.renewableenergyworld.com/rea/news/article/2010/05/taking-grid-energy-storage-to-the-edge, "Taking Grid Energy Storage to the Edge," an article by Brad Roberts, S&C Electric.

The term community energy storage (CES) defines an approach where smaller packages of battery energy storage, typically 25 kW with one to two hours of back-up time, are deployed in neighborhoods on street corners or along backyard utility rights-of-way.

The CES units are connected on the low-voltage side of the utility transformer and protect the final 120/240-volt circuits to individual customers. Placing a utility-controlled device at the edge of the grid allows for the ultimate in voltage control and service reliability. Meeting this challenge of even greater control of voltage at the point of customer use is a major departure for traditional utility system control philosophy, but it's needed to deal with a rapidly changing customer load profile. While customers are adding more sophisticated electronic loads (computers, appliances, etc.) requiring greater service reliability, new, even larger loads—such as plug-in hybrid electric vehicle (PHEV) charging units—will be added randomly in the grid.

[3] The Nissan Leaf is a 3.3-kW load. Industry averages for PHEVs and BEVs of 14 kWh and 22 kWh batteries result in recharge times of 4.25 and 8 hours, respectively.

[4] From the Galvin Electricity Initiative: Microgrids are modern, small-scale versions of the centralized electricity system. They achieve specific local goals, such as reliability, carbon emission reduction, diversification of energy sources, and cost reduction, established by the community being served. Like the bulk power grid, smart microgrids generate, distribute, and regulate the flow of electricity to consumers, but do so locally.

[5] A smart appliance is one that is capable of modifying its energy consumption based on price (or other) signals from the utility. It has an embedded microprocessor that enables it to have a two-way communication with the utility, sending information on energy consumption and cost. Further, the integration of the various smart appliances into a home area network (HAN) as illustrated in Figure 5.4, enables greater monitoring, control, and optimization of end-use load.

6

Business of System Operations

6.1 Anatomy of a Utility

An electric utility is in the primary business of generating power (or buying it), transmitting it to locations where it is needed and delivering power to the customer. Generation, transmission, and distribution tend to be large asset intensive areas of operation requiring large amounts of capital spending and management. As explained earlier, the utility spends money to build the assets (subject to regulatory approval) and recovers the money from the customer proportional to the amount of energy consumed at the premise.

The core operations (see Figure 6.1) of a utility are generally spread around its main value chain that generally includes generation, transmission, and distribution (also known as T&D), and customer. Other functions, such as enterprise management and the enabling functions, while still being extremely important to the proper functioning of the utility are all supportive in nature.

While this chapter will focus on the T&D in general and system operations in specific, it is important to point to few key functions in a utility that tend to stand out whether they are at the enterprise level, enabling level, or at the core functionality level. They are IT, capital management and regulatory relations, and associated compliance monitoring.

- *IT.* The typical utility is extremely IT-intensive with a unique combination of normal IT functions (like finance, ERP, HR/benefits) integrated with real-time operational systems (like EMS, DMS, SCADA) integrated with generation functions and energy trading functions. Regulations like FERC orders (e.g., 888/889) also bring in constraints such as who

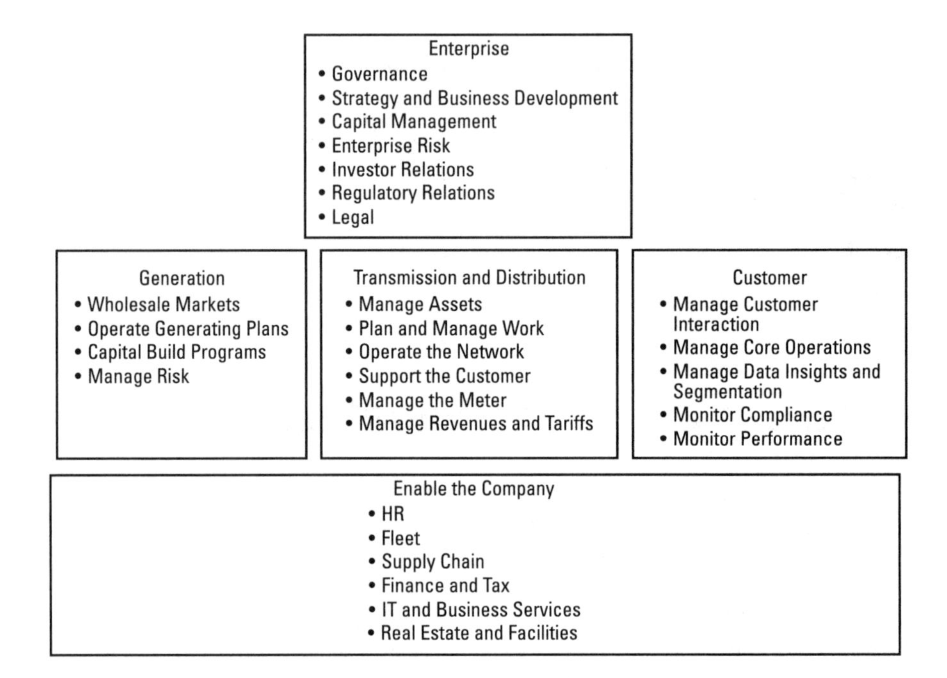

Figure 6.1 Core operations of a utility.

can get access to what information. All of this represents at best, an extremely complex IT scenario. This also brings an extra level of conservatism that tends to bubble to the surface as utilities make changes and/or replace them.

- *Capital management.* As we have mentioned before, utilities tend to extremely capital-intensive companies. Building generating plants, hydroelectric dams, transmission and distribution lines, and substations is an expensive proposition that results in asset/capital budgets at utilities that tend to be in the billions of dollars annually. Managing these large capital budgets is a complex process in every utility.
- *Regulatory relations and compliance monitoring.* The capital budgets for the most part are approved through rate cases either with the state PUCs or FERC. As a result, this is another important core capability within a utility and not only exists at the corporate level but also at the core functionality level—generation, T&D, and the customer.

6.1.1 Generation Business

Generation, as we know, deregulated due to the release of FERC Orders 888 and 889. As a result, even if the utility is vertically integrated, there is expected to be a solid separation between generation and the rest of the utility. The generation business also represents a utility's entry into the wholesale business. Core capabilities of a utility's generation business include:

- *Operating generating plants.* Operating each generating plant can take up to several hundred people with varying expertise from generation operations to journeyman/lineman. These plants are of varying complexity from small combustion turbines (CTs), which could be unmanned and remotely operated, to large coal and nuclear-fired plants, which could involve complex operational and startup/shutdown requirements.
- *Wholesale markets.* Generating plants offer their energy in the wholesale market and are thus exposed to vagaries of energy market fluctuations. To participate in these markets, they require access to physical and financial hedging instruments and also the ability to play integrate and settle the market.
- *Capital build programs.* As explained earlier, generating plants can be large multiyear, multiphased activities that require a deep understanding of the capital/funding process both within the utility and externally.

6.1.2 Customer

The customer is the only reason for the existence of the utility. The customer is critical to the very existence of the utility, since this is the person who consumes the product —electricity—created by the utility. The utility until now has held a monopoly on the relationship with the customer [1]. This monopoly means that to some extent of a jurisdiction, the traditional utility is the only entity that can supply power to a customer's premise. The customer is bound by this relationship and hence purchases electricity from the utility and pays his or her bills on time. The utility for its part ensures power is delivered to the customer in a reliable manner with minimal outages. The key functions in the customer side of the business include:

- *Managing customer interaction.* Managing a call center is one of the most important parts of this capability. This capability is the only direct hands-on connection of the utility to the customer and represents in many ways the brand of the utility. Customers tend to call the utility for all kinds of issues associated with their bill, report outages, and so on.
- *Managing core operations.* Core operations of a utility from a customer perspective covers all the activities that lead to sending the bill to the customer. This is where the amount of the bill is calculated based on

the various variables in the customer's tariff. Meter data is handled here along with its storage.

- *Manage customer programs.* In any utility, there are always different programs associated with energy efficiency, demand response, social outreach, and so on. They need to be managed and their impacts measured and validated.
- *Managing data insights and segmentation.* The large amounts of data being stored in one place allows the customer capability to perform various kinds of analysis on the data. As the utilities enter the smart grid era, they are also looking at this data, cross-referencing the data to other demographic data that they can purchase to segment their customers appropriately and service them better.

6.1.3 Transmission and Distribution

Over the last few years, while other areas of the utility like generation have also seen changes, the T&D area has demonstrated high levels of activity with a lot of changes. Everything from deregulation to smart grid has impacted the T&D area in major ways. The T&D area is also one of the businesses in a utility that tends to employ the largest number of people and so is one of the most expensive businesses to run. As a result, much of the changes that have happened here are the introduction of more automation both in the field as well as in the utility.

6.1.4 Anatomy of an RTO

RTOs and ISOs, which are responsible for the overall management of the energy market, regional reliability, and now even for coordination of the system expansion plans, have a lot of similarities with regular utilities and key capabilities that are described in this chapter. However, as we saw in Chapter 4, RTOs and ISOs in general do not own any power system assets that will drive the need for large planning departments either in transmission or the distribution arena. The assets generally stay with the asset owner and do not transfer over. This aspect coupled with no direct SCADA control of the assets drives a different kind of management and business capability structure from that of a typical investor-owned utility. Many of these differences will become more obvious in the following section.

6.2 T&D Operating Model

Even though the overall T&D capability has three main functions, they are at times divided into several subfunctions. The three main functions include:

- System operations;
- Asset management;
- Work and resource management.

In this section, we present an electric T&D operating model that will demonstrate the interrelationships between these functions. The system operations function included in Figure 6.2 will be provided in more detail in Section 6.3

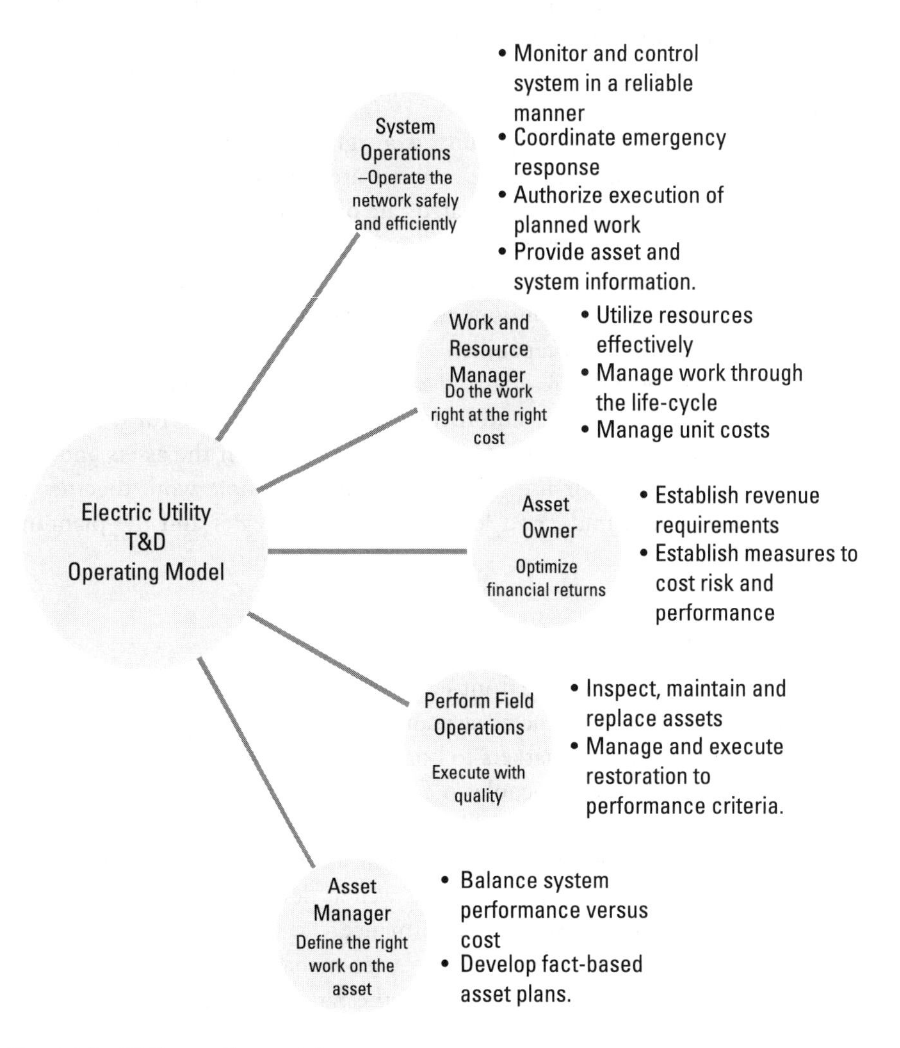

Figure 6.2 T&D operating model.

6.2.1 Asset Management

Asset management focuses on defining the long- and short-term T&D asset strategy, defining work to be executed on the assets that systematically consider the needs of all stakeholders and also integrate this strategy with the business' financial strategy (impact on ROR, free cash flow, etc.). Asset management also serves the asset owner by prioritizing capital and operations and maintenance (O&M) spending on the business through developing and monitoring strategic and operational asset KPIs [2] to monitor system performance. A GIS system tends to be one of the more important tools in their arsenal that needs to become the fundamental repository of all asset information in the utility—what type of assets it is, what is its ID, where is its location, what are its characteristics, who is it connected to, and the rendering on how the connection is made in the field (as built).

The asset management capability is a significant core capability of the utility. As we have identified before, utilities are fundamentally asset managers and are focused on getting a regulated rate of return of their investment on the assets. The asset management capability is where that happens. The entire capital plan in the utility starts and ends with the asset manager. The entire planning (transmission, distribution, relaying and protection, automation) department resides within this capability.

The system operator provides the asset manager with operational performance information and smart automation requirements. In return, the asset manager provides to system operations the information on the assets and also the policies governing their use. The two groups of people work together in developing the short-, mid-, and long-term asset strategies and the planning aspects of T&D.

6.2.2 Asset Owner

The asset owner capability is the front-end of the asset manager focusing mainly on the ownership aspects of the asset management. In this capability, the utility is going into the capital markets to borrow money to make the investment, working with the regulatory people to ensure that it is indeed covered in the rate case for recovery and finally making sure of the depreciation aspects are appropriately covered in the financial books. The asset owner also establishes the revenue requirement for the investment as well as developing the performance measures to measure cost, risk, and performance.

This is an important aspect of the asset management function in that the flow of capital is appropriately tied to the rate case, ensuring the utility's solvency in the long run.

6.2.3 Work and Resource Management

The utility's assets (see Figure 6.3) will need to be maintained on an ongoing basis, replaced when they are due for replacement and repaired when they are damaged (e.g., during a storm). Work and resource management's primary focus is to manage all of the processes associated with the assets. Here the utility manages all resources (labor, equipment, materials) required to effectively execute defined work on the T&D system. They also schedule and dispatch the work by mapping the best resources to perform the work. They perform this work by optimizing the utilization of resources to perform short-cycle and long-cycle work and also driving the overall resource forecast and plan, thereby efficiently managing work through the entire "production line" life cycle, including production design and estimating.

The large number of assets that need to be managed (across generation, transmission, and distribution) leads to a need for a large staffing pool of different people with different competencies and locational access. This capability needs to be able to allocate the right kinds of people with the right skill sets at the right location at the right time. This is because work sometimes needs to be done quickly and it is not always possible to move people with the right skill sets across large distances to fix a problem. This leads to a need to optimize the use of the right resources supported by the right processes to ensure that quality at the lowest cost is being delivered.

The system operator receives clearance requests and work schedules from work and resources management. In return, the system operator provides the necessary clearances (could include the implementation of switching orders) on the equipment so that work can be performed on the equipment. In addition,

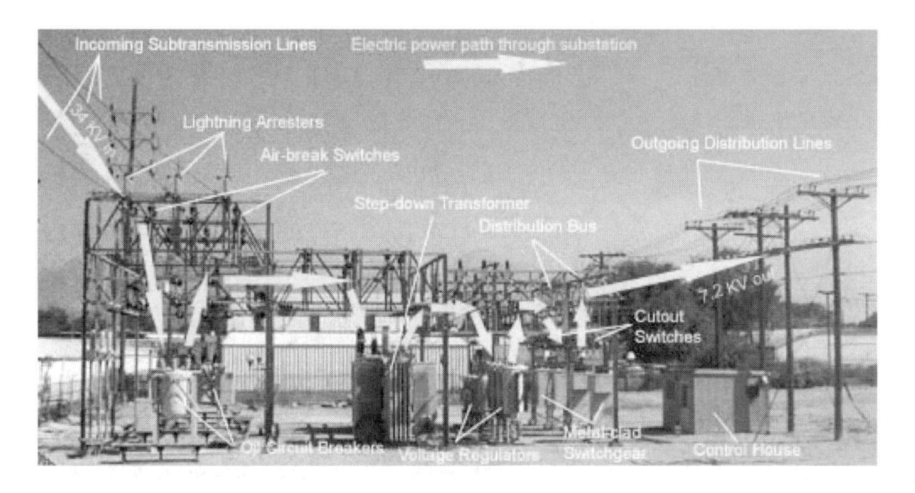

Figure 6.3 Substation equipment. (Picture courtesy of www.osha.gov.)

the system operator may also make work requests based on either outages or analysis leading to potentially imminent outages. As a result, this area works very closely with the system operator on a daily/hourly basis.

6.2.4 Field Execution

The field crews execute assigned work on the T&D system safely, productively, and with quality (see Figure 6.4) . They perform the work both during normal asset maintenance/replacement and during emergency/trouble condition. At all times, they will restore T&D network service to specified performance criteria seamlessly with internal and external resources. They will also close out work from a work quality, asset information, and accounting information perspective. There are three main types of field crews:

- *Troublemen (or T-men).* Troublemen [3] tend to be first line of attack during or after a storm and who will very often be the first on the scene and will assess the damage, call fire/police if necessary, and enter the first work order for the work to be performed along with the estimates of time to repair and the materials needed to perform the work. If the work needed to be performed is reasonably straightforward, they may also perform the work.

Figure 6.4 Utility bucket crew in action. (Picture courtesy of www.osha.gov.)

- *Field specialists.* As the name implies, specialists are capable of resolving problems in specific equipment and are generally considered utility-wide experts for that system.
- *Field bucket crew.* The bucket crews ride on bucket trucks and are literally the frontline of the utility's response to asset problems or maintenance/replacements. These are the people who do the main amount of the work.

Field crews are the first and last line of attack for a utility and its assets—they keep the system up and running. Field crews directly interact with the system operator, first to go through the switching order to confirm that the equipment is completely disconnected electrically, and visually from the system so that they can work on it. They will also work with the operator to reconnect the equipment back in when the work is completed. This handoff process is critical because of safety issues.

6.3 Architecting the Business of System Operations

The system operator [4] operates the T&D network safely and efficiently serving as the "traffic cop" for all work performed on the network (see Figure 6.5). The operators monitor/control the T&D power system, manage/maintain system reliability, and authorize execution of planned work on assets. During emergencies, the system operator is the source of all information tracked through the outage management system where it identifies outages, prioritize them, and coordinates systemwide emergency response. The transmission predominantly (and in some cases he distribution system operator as well) manages ISO/RTO interactions and interfaces and is also the implementer of any control actions needed by the RTO operator [5]. The system operator supports execution of planned/emergency work by developing switching orders to deenergize/energize the power system where needed. Lastly, the system operator is also the singular source of critical system and asset performance information

6.3.1 Drivers

In addition to the drivers that have been identified in Section 6.2.2, there are a few other timely drivers that impact the system operations space:

- Utilities will face a high percentage of operators who will retire in the next 5+ years. It takes operators about 2+ years of training before they can become proficient in their job.

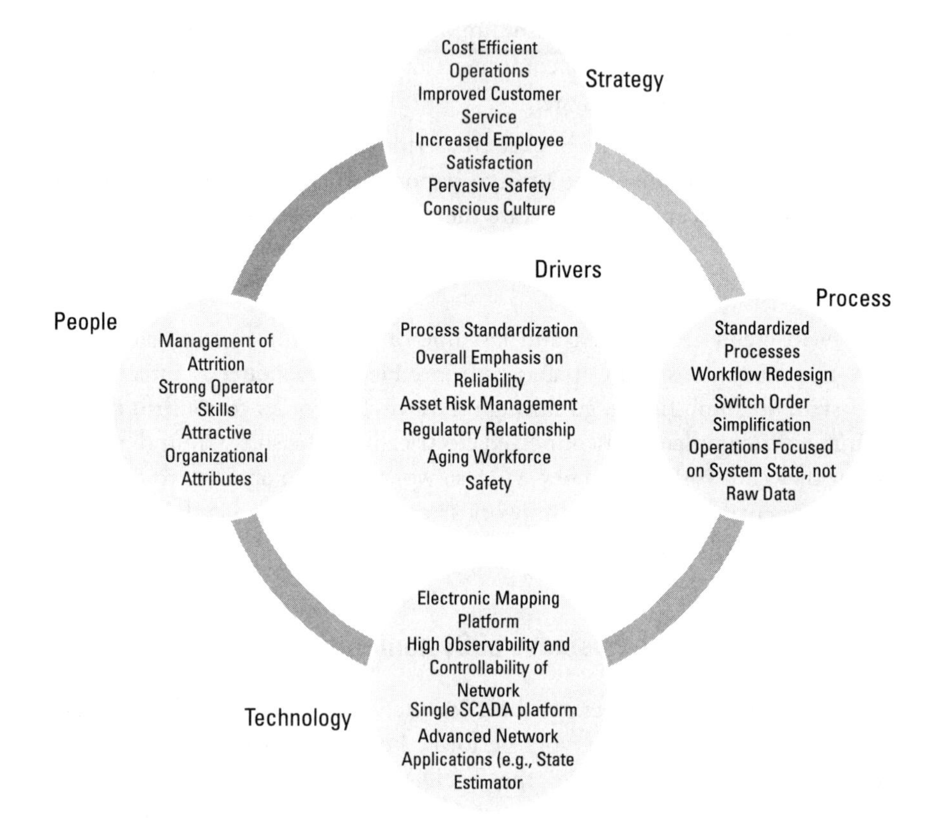

Figure 6.5 Key drivers of the system operation function at a utility.

- Utilities have increased their spending on infrastructure, and this infrastructure will now be asked to do more with less. The grid will be expected to become smarter.
- There will be greater emphasis on reliability for utilities with the creation of the Electricity Reliability Organization [6], leading to the potential for fines, some reaching as high as one million dollars per day.
- There is a greater focus on the customer for system operations when it comes to one of its core capabilities: outage management.
- There is an increased need to be operationally efficient in all parts of the utility business, and system operations is no exception.

System operations also needs to change to address these business drivers.

6.3.2 Strategy

The core strategy for systems operations has not changed much over the last few years—items like reliability and safety have always and will always continue to be critical from a strategy perspective. As the system operations arena has moved from the original vertically integrated utilities through deregulation and the Smart Grid, a new set of strategies have come into play that bring more light into the control center:

- *Pervasive safety conscious culture.* As distribution control centers move out of out the shadows of the utility service centers and DMS systems become more prevalent, the concept of safety consciousness becomes more and more important. This is further exacerbated with more devices entering the home on the premise side of the meters as well as on the grid.
- *Improved customer service.* With this move towards markets in the deregulated space and the advent of the smart meter in the smart grid space, the system operations space is becoming more opened to the external stakeholders. The external stakeholders that get access to the control center include other utilities and large customers all the way down to the individual residential customer who has a smart meter. Talking to customers means a different focus on handling the questions and responding in a professional manner.
- *Increased employee satisfaction.* The complexity of system operations means that the operator is a highly trained individual on whom the utility has spent a lot of time and money. Employee satisfaction has and always will be important to this area and is becoming more important as operations becomesmore complex.
- *Cost-efficient operations.* As with any other business, there is always a downward pressure on costs and the need to be more efficient. This is resulting in a greater need for more automation in the control center supported by advanced visualization tools and data mining and analytics.

6.3.3 People

System operations for the most part tends to be a 24/7 operation. This means that the control center operates 24 hours a day, 365 days of the year. As a result, this is one of the few areas in a utility that attracts a different kind of person—someone who is extremely motivated, knowledgeable about how the system operates and behaves, and most important, is very passionate about keeping the lights on. For the most part, the control center job has always been the most senior of the jobs in the utility's union cadre and as a result the average age of an operator is mid-fifties or higher [7]. Key areas of focus of a system operations area include:

- *Management of attrition.* As mentioned above, the average age of a system operator is mid-fifties. This means that in the next 10+ years, many of them will retire. Given the fact that these workers have come up the ranks as linemen means that they are also some of the most knowledgeable people in the utility about how the system will react under various situations. Transitioning their knowledge to the next generation of the operator will be a critical area of focus for every utility.
- *Knowledge transition.* Utility operators see situations that many times happen once in several years and may not happen again for a while. Transitioning this knowledge to other operators who have not faced this situation and to other future operators is an important area to address. Management at most utilities is actively analyzing knowledge management systems which can help in capturing the overall knowledge and store it for future learning.
- *Work under duress.* The control center environment (as is similar in control centers for other areas as well) goes through long periods of low and steady activity separated by short bursts (that can sometimes last a few days) of hectic and stressful activity, generally during storms or large unplanned outages. When this happens, the center is a very stressful area to work in and the personnel need to stay very calm as they coordinate the restoration of the system in conjunction with field crews, which in a large outage can involve hundreds of utility personnel, some of whom may even be from different utilities who may have come to help out.
- *Training.* Training is an extremely important part of becoming a system operator. Utilities spend a lot of money and effort to train these system operators, using tools like dispatcher training simulators. The training program can be anywhere from 1 year to even 4 years long before the operator is allowed to drive the operation of the system independently. Training is provided on a variety of subjects—operations processes, use of tools under normal and emergency situations, understanding the behavior of the power system under normal and emergency situations, and so on.
- *Tools.* System operators tend to work under immense pressure during system emergencies. Even under normal operations the operator tends to do a lot of things in parallel, working with several field crews in performing normal maintenance and creating/executing switching orders

to either energize or deenergize different equipment. Having the right set of tools is critical to keep track of various actions so that the sequence of actions is completed correctly and safely.

- *Work in teams.* More than anyone else in a utility, system operators work in teams. Each shift consists of a group of people who will work together to keep the power system safe and secure. Working in teams requires each team member to be an expert in one or another set of activities and depend on other operators as well as field personnel to complete specific tasks.

6.3.4 Process

For the longest period of time, system operations depended on the operators defining the processes that over time resulted in processes that differed from operator to operator and control center to control center even within the same utility. As a result the same effort would be done in different ways, preventing the utility from learning from either best or bad practices. Recently, most utilities have moved the bar to make the control center more of a process-centric organization. Given that, there is a greater focus on

- *Standardized processes.* Focusing on standardized processes has become a key priority at most utility control centers. This action has benefits not just within the control center but also with the interactions with various other departments such as asset management/system planning, work/resource management, and field crews. A specific set of definition of these standardized processes is provided in the next section.
- *Workflow redesign.* System operations are not a silo unto itself. Regardless of which process is being executed, there is a lot of interaction with people inside the control center and other parts of the utility as well as the field. Some of the tasks are completed over the period of a year (or more). This generally covers system planning tasks. Other tasks like switching sequence execution need to be completed within a day or a period of a few hours. Lastly, in an emergency situation, everything needs to be done faster and better. Given all this there is an increased interest in redesigning the entire workflow from end-to-end that will result in changes so everyone is more prepared when the work comes to them.
- *Switching order standardization and simplification.* Under normal operating scenarios, in addition to monitoring and operating the grid, the main activity in a control center is supporting the normal maintenance tasks of the power system. Switching orders are created to ensure the

equipment being maintained is deenergized in the proper sequence prior to maintenance activity and also reenergized after the maintenance activity is completed. Standardizing these procedures is crucial to the proper interaction between the various teams and also to support the workflow redesign efforts.

- *Operations focused on system state, not raw data.* System operators tend to constantly monitor the system looking for issues. A significant portion of their information comes from the alarms that are created by various items that go out of normal range and are flagged by the system. However, as the system gets more and more complex, the operator can easily get swamped under the burst of alarms. Dues to these kinds of conditions, it is important for operators to move from reacting to raw data (like individual alarms) to reacting to full knowledge of the system state. This change allows them to move from trying to find the root cause of the problem to knowing the root cause and working to solve the problem.
- *Increased use of technology.* Increased complexity of the grid activity, the sensors and controls being added to it, and the addition of renewable sources all are contributing to a serious need for strong technology-based support for the system operator's tasks. As a result, systems like EMS, DMS, OMS, GIS, and so on are all being developed and rolled out in support of the operator's tasks. Some of these tools have over time become an indispensable part of the operator's tool belt, and are also becoming more sophisticated over time with the addition of new applications.

6.3.5 Technology

Technology is the major backbone of system operations as is the case with any other critical function in a utility. More so here, the technology components used in a control center focuses on certain key attributes to support varied nature of actions that take place there.

- *Electronic mapping platform.* The mapboard or its electronic equivalent is a critical part of the system operator's toolbox. It brings a visualization aspect to the system, allowing the operator to either visualize the state of the system or look at trends at how the system is undergoing responses to stimuli put in place by the operator. The platform differs from control center to control center and also from transmission to distribution. The electronic map is slowly becoming the centerpiece of the system operator's toolbox and also becomes an excellent platform to demonstrate key actions to senior management in the utility.

- *High observability and controllability of network.* Technology in the field is a key part of the repertoire for the operator. These are basically the eyes, ears, and arms of the system operator. Sensors such as PTs, CTs, current, voltage, and PMUs all create an ability to bring data back into the control center. Controls such as the ability to turn devices and switches on and off allow the operator to reroute power or turn equipment off in order to better manage the power system.
- *Advanced suite of specialized applications.* Over a period of time, the control center has been the recipient of several sets of advanced applications. Examples like the network advanced applications (state estimator, contingency analysis, optimal power flow), generation applications (economic dispatch, unit commitment, security constrained unit commitment), and distribution applications (fault location isolation and service restoration, volt-VAR control, and feeder load balancing) all have contributed to creating a truly high-tech environment in the control center in support of the operator's tasks.
- *Integration between transmission and distribution.* A key area in which the utilities are beginning to move in the integration of transmission and distribution systems. This is still somewhat in its nascent stage given the low levels of sophistication of the distribution control center in general. However, as new systems are being procured and implemented, this integration is a cornerstone of the implementation process.

6.4 System Operations Processes

System operations processes are round-the-clock operations. As defined earlier, these processes are also not executed in a silo but are quite heavily integrated with the other core processes in a T&D organization and external. Figure 6.6 has identified seven core processes that are considered as a part of the system operator's activities. Considering the unique attributes of the system operator, these processes are also not batch processes but on the other hand are ongoing processes that are in continuous action all the time. Many are also a combination of predictive and/or reactive in nature, meaning that the system operato, while trying to predict key problems from happening on the grid, sometimes also needs to react to key events that have happened and have impacted the grid in some form or another.

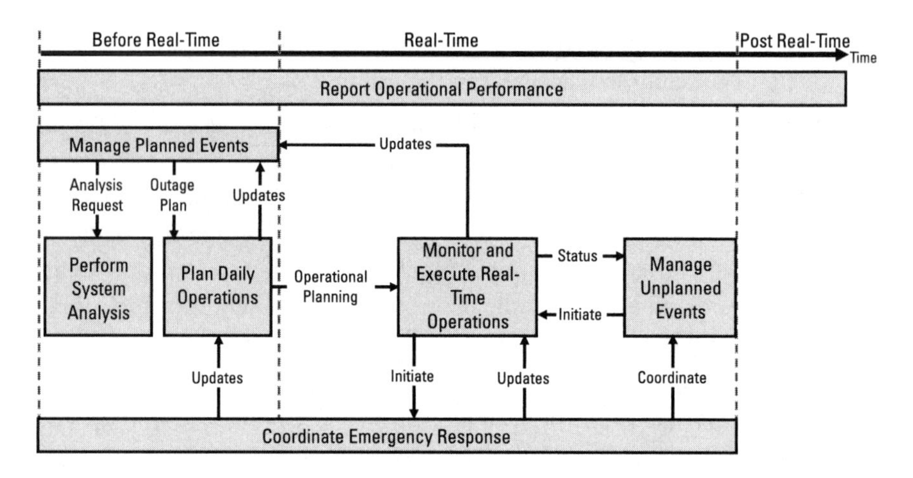

Figure 6.6 Key system operations processes over time.

6.4.1 Monitor and Execute Real-Time Operations

The most fundamental job of the system operator is to monitor and operate the grid. In this process, the system operator is constantly monitoring the state of the grid through different alarms that are coming in and watching the power flows on various lines and substations, voltage, and current levels at various points in the system. The operator is also monitoring other aspects of the power system such as generator status, system frequency, and the power that is scheduled to be transmitted through the system. All in all, the entire focus of this process is to predict (to the extent possible) potential problems that may happen and take advanced actions to prevent a problem. Key aspects of this process include:

- Ensuring the safe, efficient, and reliable operation of the electric system;
- Providing timely updates and communications to all parts of the company and external agencies (if necessary) on the state of the system and key problems;
- Minimizing response time to events impacting the system by predicting to the extent possible ahead of time and reacting in the fastest manner to resolve problems;
- Streamlining the interaction with tools used for real-time operations all across the horizon of transmission to distribution and also interacting with external agencies like adjoining utility control centers and power market operators if necessary.

6.4.2 Manage Planned Events

Any time the utility wants to maintain a specific piece of equipment, the system operator needs to get involved. First the equipment needs to be cleared for maintenance, meaning that the system conditions are OK for the specific equipment to be in a deenergized state for an extended period of time until the maintenance action is completed. The system operator also needs to actually deenergize the equipment so that the field crew can perform their maintenance actions under safe conditions. Key aspects of this process include:

- Developing and executing switching orders, which are the list/sequence of steps required to deenergize the equipment prior to maintenance and then reenergize it again after the maintenance action is completed. The exact sequence of steps is important to ensure that the equipment is deenergized (or reenergized) with minimal impact to other aspects of a stable system.
- Executing the clearance management/approval process, is one of the steps in a maintenance plan for a piece of equipment, resulting in the approval from the system operator to the asset manager confirming that the planned outage can be allowed.

6.4.3 Manage Unplanned Events

The power system is an active system in which loads are changing on an ongoing basis—loads are going up and down, generation supply is going up and down, lines get overloaded and sometimes trip (get disconnected from the system), weather-related issues could result in trees (or something else) falling on a line and tripping it, and so on. These kinds of actions are called unplanned events. When they happen because of situations beyond the control of anyone, the operators (and other utility personnel) need to react to the events and bring the system back to normal. Key aspects of this process include:

- Performing timely and effective restoration is an important part of the system operator's job. The system operator knows how to prioritize the various actions necessary to take to solve the one or more unplanned events that have taken place. Based on the prioritization, the actions taken will focus on restoring the power as soon as possible.
- Minimizing the impact of customer outage is fundamental to the operator's focus, leading them to start with prioritization of the events as well as developing the sequence of steps necessary to restore power.

- Ensuring the safety of employees and customers during the process of restoration drives the operator to also ensure that the equipment is appropriately deenergized before the field crew gets to reconnect the equipment. This same action is also applied to customers—they are asked to call the police or the utility before taking any steps on their own.

6.4.4 Coordinate Emergency Response

At some point in time, when the number of unplanned events becomes too many or the outage extent is too widespread, the utility goes into emergency management. When that happens, an entirely new organization swings into action, depending on whether this was a transmission-related or distribution-related emergency. Emergency coordination committees take over and people from different groups across the utility are seconded into this ad hoc organization. An important part of this process is the involvement of the system operator, who needs to be able to drive the reenergization process to ensure a safe and orderly restoration of power. It is not uncommon for one of the emergency center locations to be in the conference rooms adjoining the control center.

The prioritization of the restoration is done by the emergency coordination committees based on a variety of factors like criticality of load (such as hospitals), extent of outage (number of customers that can be restored with one solution), nested outages, and so on.

6.4.5 Plan Daily Operations

Every day, either by the night shift of the previous day or when the morning shift just comes in, the control center team creates the daily operations plan. This plan focuses on the day that is ahead for the team and focuses on getting ready for the action. Key aspects of this process include:

- Understanding recent system events that have just happened in the power system is an important first step in this process because it could impact scheduled work that was planned to either start or continue during the day. If there is an impact, the work and resources management processes are informed of the changes and a new updated plan is developed. This update could either be a modification of the scheduled work or a complete postponement.
- Anticipating major events like weather or special events like the Super Bowl or a major concert that could impact system integrity due to either a different load pattern and/or outages. If, for example, a new weather pattern is emerging, the system operator will appropriately modify the operations plan to prepare a contingency plan and get ready ahead of

time to the extent possible. This also requires a continuous monitoring of the operating plan to ensure that all changes are reflected into a single plan (playbook) from which everyone is operating.

- Reviewing operational plan and ensure compliance standards are met.

6.4.6 Perform System Analysis

As asset management develops their systems plan whether it is for upgrades or for maintenance of the assets, system operations is the final arbiter of the situation on whether the clearance can be provided to perform the maintenance/change. These analyses are done on different time horizons—5 years, 1 year, 6 months, next month, tomorrow, and so on. Key aspects of this process include:

- System operations will perform extensive scenario/contingency analysis of planned work scope to ensure that the system conditions will still leave it in a stable operating mode from an operations perspective. All of these analyses are preformed to ensure that the reliability of the system is not compromised by the work planned in any form whatsoever. If reliability of the system is threatened, and then they will suggest workarounds or splitting the work to still support it.
- The system operator also needs to be certain that established standards and procedures are followed both with regard to the equipment that is being maintained as well as to the integrity of the system when it comes back online.
- System operations will also look for opportunities to combine and coordinate other work that can be performed under the same clearance, thereby increasing the benefit to the utility and the ratepayer.

6.4.7 Report Operational Performance

There is a significant focus on operational performance of the power system and this is increasing with the advent of the smart grid and also due to greater customer involvement. While the focus for transmission is very different from distribution—different kinds of data are collected and reported on an ongoing basis. While the actual selection of metrics to be reported on is large and varies from utility to utility, we will discuss a small set of the industry standard ones here.

At the transmission level the main metrics are the NERC CPS1, CPS2 and DCS:

- CPS1 measures control performance by comparing how well a control area's ACE [8] performs in conjunction with the frequency error of the interconnection. CPS1 links system reliability—measured in terms of the acceptable frequency error—to the allowable control performance.
- CPS2 is designed to limit a control area's [9] unscheduled power flows. CPS2 is similar to the old A2 criteria. Average ACE for each 10-minute period must be within limits.
- A disturbance is defined as any event that is 80% of the magnitude of the control area's most severe single contingency. A control area is responsible for recovering from a disturbance within 10 minutes by recovering the amount of the disturbance or returning ACE to zero. A disturbance is not reportable if it is greater than the control area's most severe contingency. Control areas must comply with the DCS 100% of the time. Any control area not complying will be required to carry additional contingency reserve. Extra reserves must be carried for the quarter following the quarter in which the noncompliance occurs.

At the distribution level the main metrics are SAIDI, SAIFI, CAIDI, and MAIFI. SAIFI, SAIDI [10], and MAIFI are some of the indices used to measure distribution system reliability.

- *SAIFI* is the average number of sustained interruptions per consumer during the year. It is the ratio of the annual number of interruptions to the number of consumers.

  SAIFI = total number of sustained interruptions in a year/total number of consumers
- *SAIDI* is the average duration of interruptions per consumers during the year. It is the ratio of the annual duration of interruptions (sustained) to the number of consumers. If duration is specified in minutes, SAIDI is given as consumer minutes.

  SAIDI = total duration of sustained interruptions in a year/total number of consumers
- *CAIFI* is the average number of interruptions for consumers who experience interruptions during the year. It is the ratio of the annual number of interruptions to the number of consumers affected by interruptions during the year. Consumer is counted only once regardless of the number of interruptions.

  CAIFI = total number of sustained interruptions in a year/total number of consumers affected

• *CAIDI* is the average duration of an interruption, calculated based on the total number of sustained interruptions in a year. It is the ratio of the total duration of interruptions to the total number of interruptions during the year.

CAIDI = total duration of sustained interruptions in a year/total number of interruptions [11]

• *MAIFI* is the average number of momentary (less than 5 minutes) interruptions per consumer during the year. It is the ratio of the annual number of momentary interruptions to the number of consumers.

MAIFI = total number of momentary interruptions in a year)/total number of consumers

Endnotes

[1] We are seeing chinks in this armor with the opening of retail competition that is happening in Texas, Pennsylvania, and a few other jurisdictions in North American and is in place in several other parts of the world.

[2] KPIs stand for key performance indicators, which are different metrics used to help evaluate the performance of an organization or system or capability and upon comparison with other similar structures can help in defining mechanisms to improve on them.

[3] Hence the term "trouble-call management systems," which was the precursor to today's outage management systems and which denoted that there was trouble in the system.

[4] The system operator should not be considered as one person or just focused on the team of operators who all sit in the control center console. In this book, the system operator is more of a capability that includes lots of people involved in make system operations work, with the system operator and many others, such as support engineers.

[5] RTO/ISOs do not own the assets and hence do not have any SCADA-level control of the devices. The actual control needs to be performed by the transmission operator upon the request/command of the RTO operator.

[6] The Electricity Reliability Organization was created through the passage of the Energy Policy Act of 2005. Its objective was to develop and enforce compliance with mandatory reliability standards in the United States. In 2006, NERC applied for and was granted this designation.

[7] Having system operators in their mid-fifties is not in itself a sign of an aging workforce. It is more of an indicator that they will retire in about 10 years. In most utilities the system operator position is a one that people get promoted into after having worked several years in other union cadre jobs, and then stay in this position until they retire.

[8] ACE is defined as the instantaneous difference between net actual and scheduled interchange, taking into account the effects of frequency bias including a correction for meter error.

[9] A control area is an electrical system bounded by interconnection metering and telemetry, capable of controlling generation to (1) balance supply and demand, (2) maintain interchange schedules with other control areas, and (3) contribute to the frequency regulation of the interconnection.

[10] SAIFI and SAIDI are the most used pair of reliability indices. A North American survey showed SAIFI figure of 1.1 (indicating 1.1 interruption/year/consumer) and SAIDI of 1.5 hours. Singapore is reported to have a SAIDI of 3 minutes.

[11] It can also be seen that CAIDI = SAIDI/SAIFI.

7

Control Center: The Hub of System Operations

The control center is the hub of system operations activity in the utility control centers may be called by different names at different utilities or may even be used for different actions depending upon whether the control center is being used for transmission operations or distribution operations at a utility or to run the entire market/reliability operations at an RTO.

Control centers are designed to support 24/7 operations and as such all aspects of these centers are oriented around this fact. Key characteristics of control centers are:

- They are hardened structures designed to handle severe weather conditions and possibly even earthquake-proof up to a certain magnitude where necessary.
- They are supported by multiple levels of redundancy in electric power (independent feeders from two different substations) supplemented by a diesel genset, supplemented by an uninterrupted power supply (UPS), and supplemented by battery storage.
- The communications needs of the control center are also designed for redundancy and hence have multiple communication feeds coming in.

- The core personnel staffing the control center (system operators or dispatchers as they are called) is also designed around the need to staff the desks on a 24/7 basis. This means that to staff each desk adequately, somewhere between 5 to 6 personnel are needed so that key areas like vacation, training, holidays, and weekends are covered.
- The technology servicing control centers is also designed for high levels of redundancy—and hence the systems are designed for individual systems/applications/hardware to fail without impacting the overall outcome of the control center. These systems are designed with dual-redundant or quad-redundant systems so that the systems can back each other up on failure.
- Control centers also tend to have large mapboard hardware driven by software systems designed to provide the system operator with an overall view of the system with identification for power flows on lines, some critical voltages and currents identified, and system frequency [1]. Mapboard systems have evolved significantly over time, changing from static displays constructed from tiles and tape supported by lightbulbs here and there to high-definition television monitors that can display any page from the control center software, thereby allowing the system to be that much more flexible.
- Strip chart recorders tend to record the change of key system quantities like tie-line flows between control areas, system frequency, voltages, and other quantities like that over time.
- Organization of work: The core part of the control center is organized around desks; each desk has one or two operators and is focused typically on one major task. The concept of a control center desk is important because given that these are 24/7 staffed roles, they will need 5+ personnel per position.

While both the transmission and distribution control centers all basically follow the same core processes, they are organized differently, meaning that the actions performed at each desk is different. The organization is based on the criticality of the functions they perform and the level of specialization required supporting those functions.

For example, in transmission, the desks are organized as transmission desk, generation desk, and so on. Distribution on the other hand is more organized as switching desk, clearance desk, and so on.

The control center at an RTO is more organized like a transmission center with the exception of added desks to support the market functions and processes.

There are several other support functions that also function in a control center. Some of them may be 24/7 and others may be either only on the day shift or run only two shifts instead of the three. These support roles include operations engineers, system analysis, real-time operating plan, and so on.

Despite maintaining several strong and constant characteristics over a period of time, control centers have also evolved over time. In the beginning, every utility had one control center for all of their electric system operations capability. This included much of today's transmission operations, generation operations, market operations to the extent any bilateral trading took place between utilities, and even distribution operations to the extent any part of the distribution system was either remotely monitored or controlled. Over time, different configurations have emerged, especially with the advent of RTOs, gencos, energy trading companies, enhanced levels of distribution sensors, and controls supported by a DMS, and so on. All of this has resulted in the functionality that used to exist in control centers to get split among several control centers, and brought along with it a need for enhanced levels of integration that at the same time supported the conflict of interest mandates enforced by FERC orders 888/889 and others that came after it.

Figure 7.1 An actual control center in action. (Picture courtesy of Alstom Grid and used with their permission. Alstom Grid retains all copyrights for this image.)

Figure 7.2 presents a conceptual view of this split that has occurred from its genesis of all functionality inside one control center to multiple control centers, with each market participant/role focusing on the needs of their business and how to interact with the others [3].

7.1 Transmission Control Center Desks

Typical transmission control centers have between 3 to 4 desks: the generation desk, transmission desk, and scheduling desk. Overseeing these three desks is the shift supervisor's desk. In addition to this there are various support desks that may not all be staffed 24/7.

7.1.1 Transmission Desk

The transmission desk is basically responsible for overall security of the bulk power system. The transmission desk runs different kinds of analysis on the transmission system, constantly looking for weak spots and constantly asking "how can I get out of this?" The system operator has several tools at his or her disposal to perform these analyses:

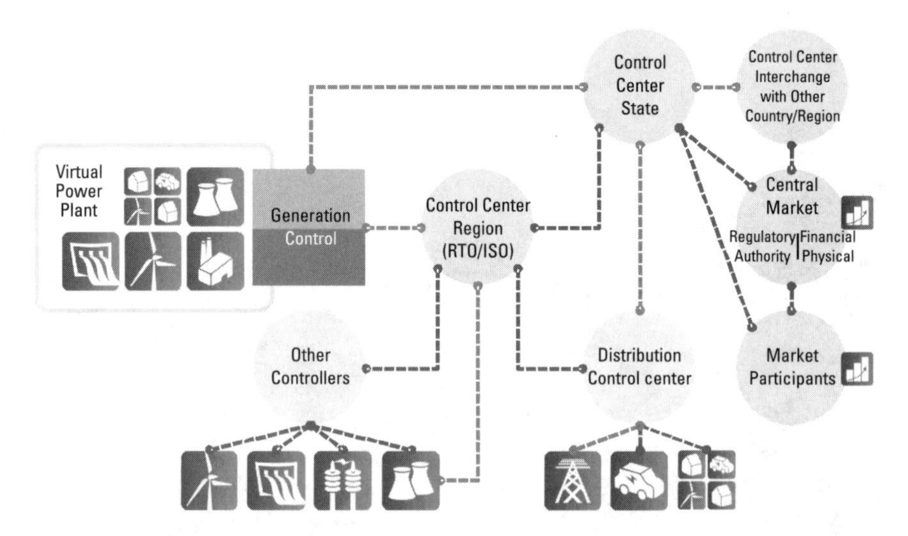

Figure 7.2 Split of original control center functionality into different market participants. (Picture courtesy of Alstom Grid and used with their permission. Alstom Grid retains all copyrights for this image.)

SCADA

This is the main application that is the eyes, ears, and arms for the operator. Through this application, sensors in the field send in data into the control center. Alarms get issued when data exceeds one or more limits that are set for them. SCADA also allows the operator to control devices in the field, such as open/close circuit breakers, change transformer taps, and so on.

It basically reports the data that it receives from the field in a raw form and converts it into engineering units based on the actual quantity that it is sensing. The data could be any of current (amps), voltages (kV), or power flow (megawatts (MW) or megavolt ampere reactive (MVAR)). It could even be something like the status of a substation door (open/close).

The incoming data from SCADA has all kinds of variables—generator output MW, MWh, MVAR, and MVARH, power flow values through lines measured either in MW and MVAR or current values in amperes. Much of this data is sensed in the field, collected by devices called RTUs [3] or IEDs [4], which then send the data to the control center through a dedicated set of communication lines.

SCADA has one main problem. It does not know if the data is accurate or not. It cannot also confirm whether the circuit breaker status is open or not, even if the data that comes in says so. These errors could happen due to the meter either reading wrong or just malfunctioning. It could also happen if the meter parameters are adjusted in error, leading to inaccurate conversion from raw to engineering units. When this happens, the information is still reported as is, and this is where the state estimator comes in.

State Estimator

The state estimator performs an estimation of the power system state. The state of the power system is defined by the combination of the voltage magnitude and the voltage angle [5] at every bus in the power system. The key to this calculation is that if we know these two variables, almost every other quantity (current, power flow, etc.) in the grid can be calculated.

The state estimator is executed generally every 5 minutes and when some component in the power system changes state (e.g., circuit breaker gets tripped). It starts with getting the data from SCADA and ends with the calculation of system state.

Once the system state is calculated, then the power system variables and the flow through various components need to be calculated. This is performed by another application called the powerflow.

Power Flow

Power flow is generally the last step of state estimator calculation. It is used to start from the system state information (which actually does not mean much to most people), and convert it into power flow calculations that then inform the operator the calculated value to power flows across various components in the power system. The result of power flow calculation includes both the real and reactive components of power flow (in MW and MVAR). This output calculated information is then superimposed on the one-line diagrams used by most operators to monitor and operate the grid. This information is also passed on to the alarms subsystem to bring the operator's attention to (1) alert them of a out-of-normal value, and/or (2) that a SCADA measurement could be wrong and needs to be fixed.

The next step is to test the power system for reliability, and this is when the contingency analysis application comes in.

Contingency Analysis Application

The contingency analysis application performs an analysis of the power system response to various potential problems (contingencies). The basic premise here is that the power system at normal state should be able to withstand a predefined set of contingencies and still stay secure. This criterion has been established by NERC [6] from a system security perspective.

The fundamental criterion is that the system should be able to withstand the loss of any power system component, such as

- Loss of a single generator, generally the largest single generator in the grid;
- Loss of any power system component (e.g., the loss of a line, a transformer, or a circuit breaker).

A single-component contingency analysis checks the systems security status for the loss of any of the components identified above—and that the system should still be secure. The contingency analysis application will run multiple power flows, one on each of these components to give the operator a better understanding [7] of what will happen to the system state if any of these components fail (see Figure 7.3).

Based on analysis, understanding, and experience with the power system, system operators also identify a set of multicontingency cases, which means that if a combination of components fail, the system will still stay secure.

Figure 7.3 A typical control center mapboard example. (Picture courtesy of Alstom Grid and used with their permission. Alstom Grid retains all copyrights for this image.)

Optimal Power Flow

If contingency analysis identifies one or more contingencies that would result in an insecure state for the power system, the optimal power flow (OPF) application looks at the best and most feasible option to move the system from a precontingency state to a postcontingency state. The OPF takes only physical constraints into consideration. Market mechanisms are not taken into consideration here.

Security Enhancement

Security enhancement is an improvement over OPF in that it takes market mechanisms into consideration to identify steps to be taken to move the system from a precontingency state to a postcontingency state.

The transmission operator tends to have the most difficult job even though it appears (at first) to be simple. Using the tools identified above, the operator is constantly looking at the systems space, looking at various one lines, looking at various substation details, and trying to understand what is happening on the system. A transmission operator needs to be constantly on the lookout—at SCADA alarms coming in, other activities in the network (there is always planned work going on), generators are coming up and down, and lines are getting loaded up and down. From a maintenance perspective, transmission lines are going in and out of service, circuit breakers and transformers are getting maintained, which means they are going in and out of service, and so on. A transmission operator watches the system on an ongoing basis.

During the night shift, since there is not much maintenance activity going on, most operators tend to use this time to take care of paperwork and other administrative tasks while monitoring the systems.

Over a period of time operators tend to get much more comfortable with the network; they have a set of indicators that they watch for and from there they can quickly figure out whether there is a problem with a segment of the network.

Transmission operators also have a feel as to the flows and the schedules of networks around them. For example, in PJM, the transmission operator knows all the normal flows of most major transmission lines in Ontario, New England, New York, ECAR, MISO, and so on because any changes in any of those flow will have impact on the PJM system.

7.1.2 Generation Desk

The generation desk is very different from the transmission desk. The generation schedules come in either a day before, or on the hour for the next 3 or 4 hours based on a preset schedule. The generation operator's main job is to look at the overall load, confirm that between schedules (generation external to the system but coming into the system due to a prearranged sale or purchase) and internal generation it is sufficient to meet the load (consumption). The total amount of generation should meet load, plus any necessary reserves and also to cover any outages that may be identifiable ahead of time. The generation desk also looks at the potential to bring on new generation in a hurry if and when an actual unplanned outage takes place.

The primary application suite at the generation desk is called the real-time generation application [8], study generation applications, and so on.

The core functions under real-time generation are economic dispatch (ED) and automatic generation control (AGC).

- Economic dispatch looks at the list of generators that are online and running and their cost models (or market bids) if there is a wholesale market. Based on this stack, ED decides which of the generators should be generating how much.
- AGC is the real-time application which tends to run every 2 to 4 seconds depending on how often SCADA runs, and then uses that information to get an understanding of exactly how much each generator is generating at the moment and what the load is expected to be in the next 2 to 4 seconds. Based on that information and the results of ED, AGC sends control signals to the generator to either ramp up or down, or do nothing. If the generator is not responding to the control signals, then an alarm is raised to the attention of the operator.

The primary application within the study generation suite is unit commitment. Unit commitment basically focuses on how to commit generating units for the slightly longer term. While ED assumes that generators are both online and running or can come up very quickly without too much preparation, unit commitment assumes that generators can be started up. This means that its time frame of reference is more of the next day, week, and so on.

Much of the focus of unit commitment in the present day and age is on market functions and the need to move away from cost to market bids for both generation and ancillary services. Taking generation availability/bids along with the ability of considering congestion has led to the movement away from a pure marginal cost across the entire system to a locational marginal cost where the price varies based on locations in the systems and their respective localized loads.

In utility control centers, the generation desk has been reduced in importance over time due to their becoming more of a market function, which is generally controlled by the RTO if one exists in the region.

- In places where an RTO exists, much of this functionality has been moved to the RTO's control center. It starts with the market functionality, which drives and bids and the market clearance activities. The bids that clear then become generation schedules (for energy and ancillary services) that are dispatched out of the RTO's generation desk to the gencos generation desk and are then tracked.
- In places where an RTO does not exist, this work is done by various gencos (either owned by the utility or other) who may buy or sell power through bilateral trades or from other exchanges, all of which get recorded as generation schedules and dispatched as such. These genera-

tion schedules get sent in to the balancing authority where they are dispatched and tracked.

However, regardless of where it exists, the generation desk is a very important function for the overall control center and also from the perspective of ensuring enough generation always exists to support the load.

7.1.3 Scheduling Desk

The scheduling desk primarily focuses on all activities that require some kind of scheduling. The primary quantities that need to be scheduled in a control center are energy and transmission access. Other scheduling activities like outage scheduling (for planned maintenance work on the system) are also tracked and scheduled but generally performed by the support desk because they do not need to be staffed 24/7.

- *Energy scheduling.* All energy coming into the system, going out of the system, or going through the system under control of the control center needs to be scheduled with the control center. The control center needs to know about all the energy flows in, out, or through the system because the generation desk needs to use this information to define the economic dispatch or unit commitment stack.

  The scheduling desk is fundamentally responsible for approving (or rejecting) the energy schedule request. The key aspects that the scheduling desk checks before approving or rejecting the request is, Does it violate any ramp [9] characteristics? Can this energy flow? If it requires any generation is there a corresponding generation schedule to back that up? If there is load then is that a real load or is it going to be wheeled out of the system again?
- *Transmission access scheduling.* Transmission access scheduling started as a direct result of FERC open access rulings 888 and 889. Under open access rules, utility companies must allow external entities fair access to the electric transmission systems in North America. The act's intent was to allow large customers (and in theory, every customer) to choose their electricity supplier and subsequently pay for the transmission to deliver it from the generation to serve their load.

  To deliver to this mandate, the scheduler is expected to evaluate every transmission access request that comes across the OASIS system in an open and transparent manner and either approve or deny the request. All approved requests are stored in the system and used by all the other desks (both generation and transmission) and the support functions.

The approval (or denial) is based on whether there is adequate transmission access available under the product (firm, nonfirm, etc.,) for which the requests are being made. Once the approval is done, the remaining available transmission capacity is then updated so that all the participants in the transmission access market can get the same information at the same time.

7.1.4 Other Support Desks

In addition to the three main desks that have been described in the sections above, there are several other desks that function within a control center, some that may be staffed 24/7 and some that may not. For the most part, these are support and engineering functions required for the control center and the rest of the utility's transmission operations to function smoothly. A representative list of support desks include:

- *Shift supervisor desk.* As defined earlier, the core control center desks are 24/7 and one or more operators staff the desks based on the volume of activity and complexity of operations. Each shift also tends to have a shift supervisor whose main role is to be the supervisor for the other operators. There will always be key decisions that need to be taken during the shift and the shift supervisor typically tends to have the authority to approve (or deny) them. Very often the shift supervisor also tends to be the management person to whom the operators report. This desk is typically also staffed on a 24/7 basis. From a location perspective, this desk is generally situated either in the middle of the other desks or behind them.
- *Clearance management* [10]. Anytime before any equipment on a power system is touched for maintenance, a clearance needs to be obtained on that equipment (see Figure 7.4). A clearance means that permission has been obtained for the utility to perform physical action on it. The request to perform maintenance (a clearance request) is sent by asset management. The clearance operator performs the analysis based on the projected system conditions for the time frame of the clearance and confirms that the system can perform in an unobstructed manner while the component is out. Once the clearance has been provided, a tag is placed both on the physical device in the field as well as in the systems used by the operator. The tag alerts the people in the field that between a certain date and time combination, work is being performed on the equipment and that it should not be energized inadvertently.

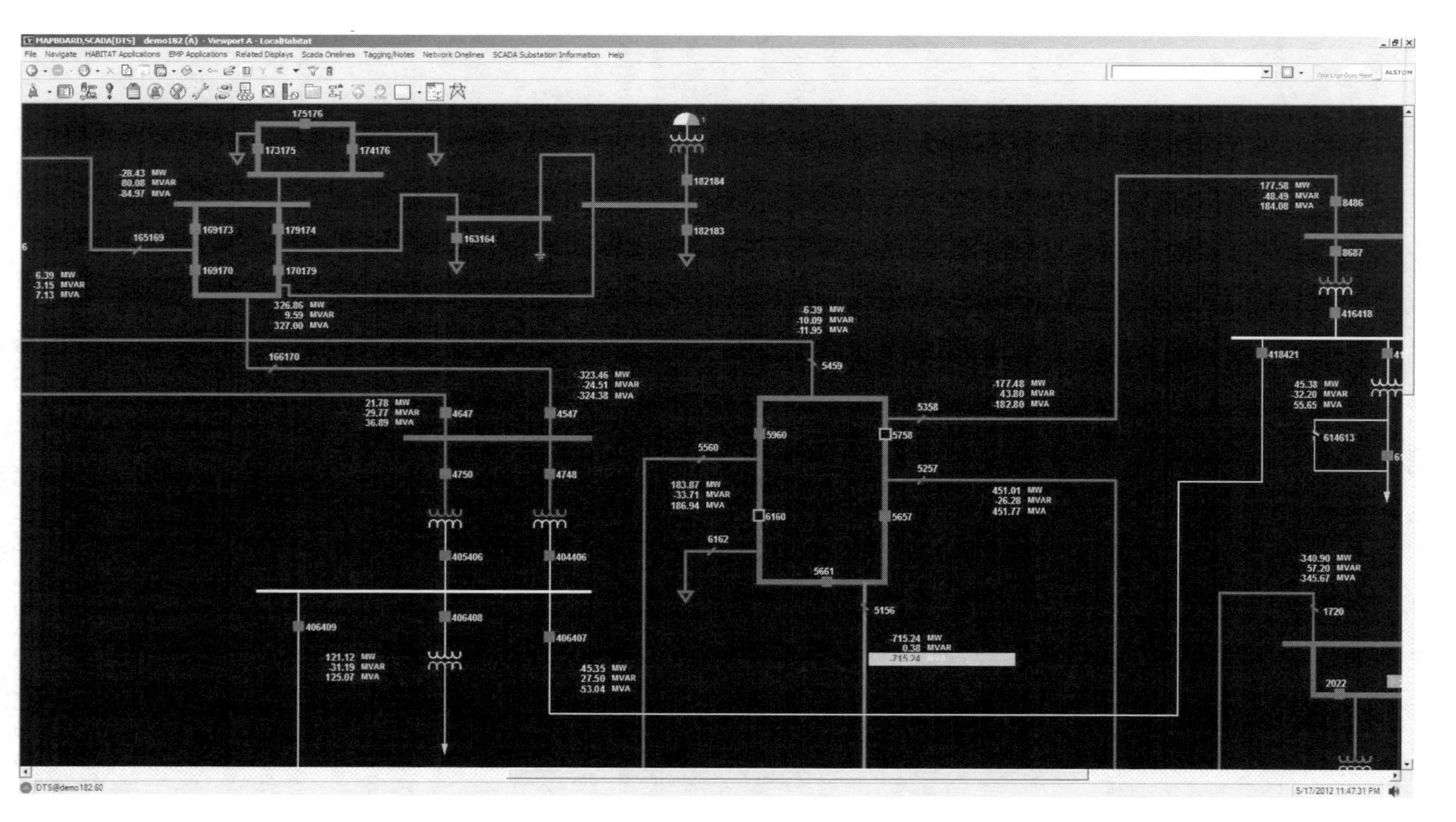

Figure 7.4 Typical transmission one-line diagram. (Picture courtesy of Alstom Grid and used with their permission. Alstom Grid retains all copyrights for this image.)

- *Switching desk.* Before any work is actually performed on the equipment, it needs to be deenergized. For this to happen, the system operator will go through a fairly complex set of steps of circuit breaker openings and closings to ensure that (1) the particular equipment of set of equipment is completely deenergized, and (2) the power system has been reconfigured in a way that the system can perform in a stable manner while this equipment is not available. The transmission desk will normally perform or coordinate the performance of the complex sequence of operations necessary to deenergize the equipment. However, developing this sequence of operations known as the switching order [11] is actually done by the switching desk in conjunction with the clearance desk. This is a very important activity because a specific piece of equipment can be deenergized in many ways. The person manning the switching desk will (based on their long years of experience) come up with the best and most efficient sequence of steps possible.
- *Operations support desk.* The operations support desk is generally staffed by operations engineers who provide a lot of support to the operators. These personnel perform short-term, mid-term, and long-term studies using different applications, some embedded within the EMS and some external to the EMS, and are constantly giving feedback to the operator, asking them to watch out for specific conditions that they may come across and also what to expect under those conditions and what key options they have for a response. These support engineers are also working on KPIs, reports (internal and external), and also generally monitoring the power system's performance to various actions. The operations engineers also maintain the power system, SCADA, and generation models for the various systems in the control center.

7.2 Distribution Control Center Desks

The distribution control center works very differently from the transmission control center. Unlike the transmission control center, the distribution control center did not get affected at all by deregulation, but is expected to be more impacted by the advent of the smart grid.

The origins of the distribution control center started from the original trouble-call management centers from where outages were being managed. Because of this core requirement, these centers were generally located in one of the back rooms of utility distribution service centers so that they could coordinate during storm (or other emergency) restoration processes by directly interacting with the field crew.

The biggest aspect of work in a distribution control centers is clearance management and developing switching orders for planned work. This is because until now, for the most part, there is very little observability into the distribution system and limited controllability as well. The large amount of planned work being done in the distribution system also results in making this area one of the more important aspects of the control center. As a result, the desks in a distribution control center are focused around clearance management and switching order creation.

Given the need for a high degree of coordination between the control center operator and the field crew, they continue to be located in services centers, where a great deal of planned work happens in the system every day (could be in the hundreds in a large utility) coordinated from multiple service centers.

Distribution operators for the most part worked using paper maps, which were the best representation of the connectivity in the field. These paper maps were maintained lovingly by the operators and updated very few times a year by the graphics people. The operators would pore over these paper maps as they developed switching orders—which came only from many years of experience. Sometimes, these paper maps were indeed connected to a rudimentary GIS system (mostly home-grown) that might be needed to support the underlying data model for an OMS.

The main area of change in a distribution control center, a charge led by the advent of the smart grid, is coming from new systems (integrated DMS/OMS), replacement of paper maps with electronic maps integrated with the DMS, and the need/ability to handle distributed generation and storage from the end of a distribution feeder.

One common trait that they share with the transmission control center is that these centers also tend to be staffed 24/7.

7.2.1 Clearance Desk

As explained earlier, this desk provides clearances for planned work to be done on the power system. This desk performs a very important task within the control center because the operator at this desk needs to analyze the loss of this equipment under different system conditions for the entire length (of time) of the clearance.

The clearance request comes from asset management who is interested in maintaining a specific piece of equipment—a necessary step to ensure the normal long-term operation of the equipment under different conditions. The clearance process is a close coordination between asset management and the distribution operator who will identify one or more specific times during which the clearance can be taken. The operator will also look at other clearance requests to see if any can be bundled into one single clearance action, thereby ensuring that planned equipment outages are minimized while at the same time supporting the needs to maintain the equipment.

As clearances are issued, there are also possibilities that they can be revoked. If the system conditions change so much that the original assumptions of system conditions are no longer valid, the clearance desk can revoke a clearance, at which point the process will start all over again.

7.2.2 Switching Desk

As explained earlier, the next step to an approved clearance is the development of a switching order. The main objective of the operator at this desk is to develop switching orders. Switching orders in distribution are very similar to their counterparts in transmission with the following exceptions:

- *There are so many of them.* Unlike transmission, where the numbers of planned work events are fewer, a lot of planned work takes place in distribution.
- *Fewer tools available for automation.* Unlike transmission, where there are several automation tools are available for the operator to test the steps out in a safe simulated mode, the distribution operator has to test hypotheses out on paper and use his or her best experience to avoid any missteps or problems.
- *Fewer options to reroute power.* Developing a switching order is more challenging because of the primarily radial nature of the distribution system and the three-phase unbalanced system. This aspect compared to the networked/grid nature of the transmission system, which is also balanced, makes the two very different problems to solve.
- *More steps in the process.* Distribution switching orders tend to have more steps in them on average and the added complexity of having more of the steps being manual in nature, which means for the execution of most of the steps, the operator will need to work with a crew in the field even to implement several of the steps in the switching order.

7.2.3 Other Support Desks

Very similar to the transmission control center, the support desks in the distribution control center are (1) the shift supervisor desk (or senior dispatcher desk) and (2) the operations engineer/analysis desk. The latter desk for the most part is not staffed 24/7 (just as in transmission) but may be staffed two out of three shifts, mainly to accommodate the volume of planned work (see Figure 7.5) [12].

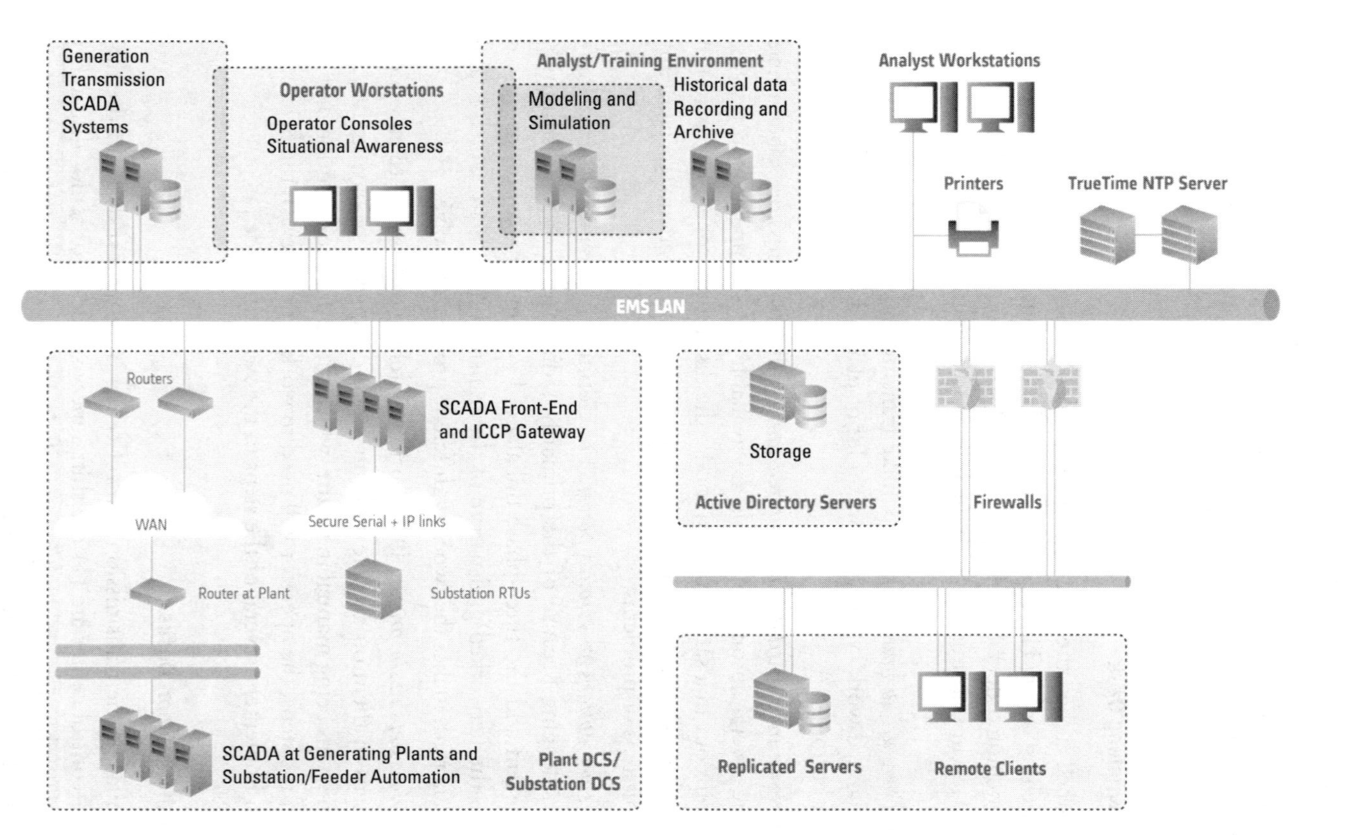

Figure 7.5 Typical architecture of all the systems in a control center. (Picture courtesy of Alstom Grid and used with their permission. Alstom Grid retains all copyrights for this image.)

7.3 Other Key Aspects of a Control Center

Control centers are differentiated from any other department in a utility due to a set of key characteristics:

- *Mapboard.* Depending upon whether this is a distribution control center or to be used for transmission control, the mapboard varies from large paper-based boards to tile-based to fully electronic displays. Regardless of the mechanism used, they clearly distinguish the control center from everything else. The mapboard also allows the operators to get an overview of the system at a glance with a focus on some of the main attributes that could be of value to all.
- *Complex telephone systems.* Control centers also tend to have complex telephone systems with advanced caller ID systems generally integrated with a wide variety of new and legacy phone, voice over Internet Protocol (VOIP), cellular, and radio-based systems.
- *Alarms and annunciators.* Control centers tend to have alarms and annunciators along one of the walls. The main idea behind this panel is to bring the operator's attention to one or more specific problems on the grid that requires immediate or quick response.
- *Audit tracking.* Control centers are the hub of the operational activity in a utility. Both during normal processes and emergency/storm restoration processes, numerous decisions get made, most of which are right and some which may be wrong. Utilities like to learn from both—which could either be used for training the operators or for performing analyses for a report. Most actions taken inside a control center are tracked for future auditing trails. This applies to the data entered into systems like EMS/DMS and so on. It also applies to all phone calls made in and out of the control center. The stored information can be retrieved at a future date/time for analyses.
- *Failure redundancy architecture.* The 24/7 nature of control center systems requires that they are architected for fault tolerance. This means that both the applications (software) and the hardware components are designed to stay up all the time and when one or more of them go down, the system is designed such that the data and models are not just secure but that the appropriate components are brought back into operation with little or no time/system access lost.
- *Physical security and cybersecurity.* The importance of system operations in a utility in part comes from the capability to open/close circuit breakers that in turn could black out large parts of their territory. As a re-

sult, a lot of attention is focused on both physical and cyber security.

From a physical security perspective, NERC has already identified key actions that need to be taken to record the ingress and exit from the control center. They generally require that every person (including visitors) sign in with full disclosure information (e.g., name and affiliation). Visitors are also escorted within the control center facility.

From a cybersecurity perspective, the systems within a control center are generally held to a higher standard of cybersecurity protection. This action results in impacts to how the systems are architected, potentially resulting in separation of how data and actions are passed in and out of the systems.

7.4 Introducing a High-Performing System Operator

System operators face the same set of pressures as the rest of the utility to continue to innovate both from a technology perspective as well as from a business perspective. All of this is a focus on moving toward a high-performing system operator.

The drivers towards this innovation are:

- *Process standardization.* There is a need to standardize all the processes that impact the system operator and how they interact with other parts of the utility as well as those external to a utility.
- *Overall emphasis on reliability.* The continued emphasis on system reliability and the public uproar after the 2003 Northeast blackout has led to new and improved mechanisms made available to the system operator.
- *Asset risk management.* At one level, the system operator's job is also about managing the asset risk in a prudent and effective manner.
- *Regulatory relationship.* With the push for the system operator to include distribution and move towards the customer, the regulator is now more interested in understanding the impact of operations.

To respond to these drivers, system operations are slowly moving in a direction of change (Table 7.1).

As this movement is occurring in many control centers around the world, there are several benefits being identified:

- Reduction in operating costs;
- Improved operator productivity;
- Improved customer service;
- Support system growth, with major growth occurring in greenfield locations;
- Improved response time during emergencies—customer satisfaction scores and reduced outage duration.

In order to support these benefits and also to provide mechanisms for control centers to improve over time, several metrics have been identified in control centers in North America and around the world. Some of these metrics are required by NERC, some by state PUCs, and some are being implemented by utilities just because they support an internal intent to continuously improve and also do more activity with either the same set (or fewer) number of people (Table 7.2).

Table 7.1
Changes in a System Operator's Responsibilities

What's Ending	What It Is Moving Towards
Control centers operating in silos with visibility of only their geographic area	Increased remote, SCADA-based controls for operating the network
Customers and emergency agencies are primary methods for determining network problems	Visibility of network status and information between control centers
Ambiguity around workforce switching skills and capability, leading to a small pool of trusted switchmen	Electronic schematics with real-time network status information
Paper- and tile-based static wall maps	Improved coordination of clearances and work with the goals of maximizing the amount of work completed with a given clearance and minimizing the number planned outages
Customers with outages experiencing long periods with inadequate information about restoration	A worker classification that is qualified to both first response and switching (or the development of a switchman classification)
Switching orders with hundreds of steps	Network operability is consistently factored into new designs
Individual workers choosing the work they are assigned	Resourcing in storms/events is done from a centralized scheduling/dispatch center
Using different tools, processes, and people to do resourcing in storms/events	Initial estimated time of restoration (ETOR) is based on historical information and operating experience
	Streamlined switching orders coupled with the expectation that switchman have the proper skills
	Consistent, standardized business processes between control centers
	Operational preparation to accept new load on the system is uniform across all control centers
	Improved integration between operations, operations engineering, and mapping

Table 7.2
Key Attributes of a High-Performing System Operator

Area	Sample Metrics
Reliability	SAIDI, CAIDI, SAIFI, MAIFI Compliance with NERC/ERO reliability and control area rules (e.g., CPS1/CPS2) Operating reserve compliance Proven backup/recovery capability System/communications uptime
Operational Efficiency	Operating costs (GMC) Actual vs. optimal dispatch Switching errors Clearance application response timelines Need something on emergency response
Information Accuracy	Forecast accuracy (long- and short-term) Transmission capacity accuracy (OASIS) Planned outage accuracy (start/stop) Procedure accuracy (e.g., switching procedures, emergency procedures) Accuracy of operator logging
Customers	Customer satisfaction survey Power quality (voltage, frequency, etc.) Billing accuracy/timeliness
Employee	Employee satisfaction survey Overtime/sick days Safety (reportable events, lost time accidents, etc.) Training (maintaining operator certification, cross training on various desks, etc.)

Endnotes

[1] Control center picture (Figure 7.1), courtesy of Alstom Grid and placed here with their permission. Alstom Grid retains all copyright for this picture.

[2] Figure 7.2, Picture Courtesy Alstom Grid. Picture placed here with their permission. All copyrights of this picture remain with Alstom Grid.

[3] An RTU is a device that collects data from data acquisition equipment and sends them to the main system over a wired or wireless network. See SCADA.

[4] An Intelligent Electronic Device (IED) is a term used in the electric power industry to describe microprocessor-based controllers of power system equipment, such as circuit breakers, transformers, and capacitor banks.

[5] A "phasor" is a complex-number representation of an electrical quantity, such as voltage, current, or impedance. These ingredients must be present in any definition of a phasor. A phasor, (possessing both magnitude and direction), is not the same as the vectors commonly used in other areas of physics (e.g., force vectors, electric/magnetic field vectors, etc).

[6] NERC's major responsibilities include developing standards for power system operation, monitoring and enforcing compliance with those standards, assessing resource adequacy, and providing educational and training resources as part of an accreditation program to ensure power system operators remain qualified and proficient. NERC also investigates

and analyzes the causes of significant power system disturbances in order to help prevent future events.

[7] Figure 7.3 is a picture of a typical control center mapboard, Courtesy Alstom Grid. Picture placed here with their permission. All copyrights of this picture remain with Alstom Grid.

[8] Real-time generation and study generation are generic terms used in the industry. Each EMS vendor uses different names for them within their system.

[9] Energy cannot just go from zero to the end result value in an instant. It needs to be ramped up and down in an orderly fashion so that the power system does not see too many volatile changes in a short instant.

[10] Figure 7.4 is a picture of a typical Transmission one-line diagram, Courtesy Alstom Grid. Picture placed here with their permission. All copyrights of this picture remain with Alstom Grid.

[11] A switching order contains a list of switching devices and time of execution for circuit breakers, load disconnects, and ground disconnects. Before any switching sequence is executed, the operator verifies whether the sequence is compliant with safety switching procedures and requests confirmation during execution of each step before proceeding to the next step in order to avoid inadvertent switching.

[12] Figure 7.5 shows a typical architecture of all the systems in a control center, Courtesy Alstom Grid. Picture placed here with their permission. All copyrights of this picture remain with Alstom Grid.

8

Energy Management Systems

The acronym EMS stands for many things these days. Even when expanded into its full form—energy management system—it can have many meanings, such as building energy management systems and so on. However, when you walk into a control center and talk about EMS, it stands only for one thing: the core flagship system used in a transmission control center that is responsible for the management and control of much of the transmission system in a utility or RTO.

> An EMS can be defined as an integrated system of computer hardware, software and firmware designed to allow a system operator in the control room to monitor, control and optimize in real-time the flow of electric power in a transmission system through the use of advanced algorithms, intelligent techniques and situational awareness-based visualization mechanisms [1].

Even within a utility, the EMS can have different meanings. Some of these have more to do with history than with the actual use and value of the systems involved. From a historical perspective, the first entrants into this arena were the SCADA systems. This came from a need to control devices remotely and get some of the sensor data into a central control center. The 1965 New York blackout led to the identification of a need for more sophistication, and the need for (what was then called advanced applications) network applications was born. This was timed well with the release of the development of foundational algorithms for power flow, optimal power flow, and so on.

With the advent of the advanced applications, the modern EMS was (kind of) born. The 1977 blackout in New York led to the identification of a need for better training for operators so that they could respond to one-of-a-kind problems if they were trained for it. The DTS also required a complete redesign of major modules of the EMS because, for the first time, the system needed to be able to either go at normal clock time, faster than clock time, and also be able to go back and forth in time so that operators could be trained multiple times on the same scenario. The simulator also needed to allow a trainer to set up scenarios that could execute events in the power system and be able to observe the response of the trainee operator.

Over a period of time, the importance of the EMS continued to grow and the data (both real-time/raw and the calculated data that was the output of different applications) that it collected was required outside the EMS and the control center. This rise in its importance required that the EMS become more and more integrated with other corporate business, analysis, and reporting systems. The continued evolution of the control center also created a need for closer integration between control centers of different utilities where real-time data needed to be shared. New protocols were then created (specifically Inter Control-Area Communication Protocol (ICCP)-based data exchange) so that control centers having EMS systems from different vendors could share data in real time.

Within the utility, this movement required closer integration with other parts of the utility and also led to change in how EMSs were architected. The new architecture led to the opening of the systems from proprietary systems into some kind of SOA-based architectures [3] (Figure 8.1).

This chapter analyses EMS from the following perspectives:

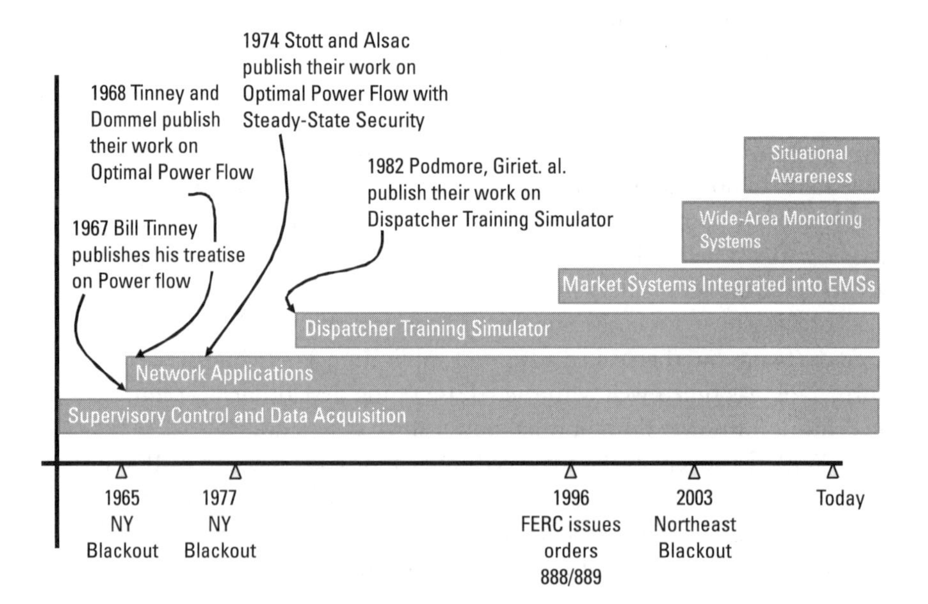

Figure 8.1 The evolution of an EMS.

- How an EMS is used to support the mandate of the system operator by different users;
- Key components of the EMS—hardware, software, databases, user interface, and power system modeling;
- EMS application suites—SCADA, network applications, generation applications, modeling applications, system utilities (alarms, process manager, and others), and WAMS.
- Other tools in the transmission control center

8.1 How an EMS Supports the System Operator's Mandate

EMS is one of the core systems supporting the system operations mandate of different entities, all of whom are responsible for the security of the electrical power system in one form or the other. However, over time this unique system has grown so much in depth as well as breadth that this name has become synonymous with the use of any real-time system in a utility control center. As a result, this system in some form or another is being utilized by the transmission operator, generation operator, RTO/ISO market operator, and even the RTO/wholesale market participant (from within the utility or independent).

8.1.1 Transmission Operator

The transmission operator is the traditional user of this system. EMS was fundamentally designed for this type of operator. As a result, the transmission operator uses this system to perform all the normal actions expected of him/her: monitor, control, and optimize in real time the flow of electric power in the transmission system. In a vertically integrated utility with no RTO, the transmission operator is also (very often) the primary responsible agency for reliability for the region. However, if an RTO exists, these commands and instructions will come from the RTO system operator to the transmission operator for implementation.

8.1.2 Generation Operator

There are two types of generation operators. One set is responsible for one generating unit and/or plant and the other set is responsible for managing an entire fleet of generating plants. The former uses a more specific process-based system called a plant DCS [4] while the latter uses a more specific configuration of an EMS and sometimes the term generation management system (GMS) is also used. While the underlying infrastructure and systems mechanisms are the same between GMS and EMS, the actual set of advanced applications may be customized to support generation operations instead of transmission.

The generation EMS will focus on monitoring the real-time statuses of the plants, and in addition

- Interact with one or more RTOs and other control areas to which they deliver power.
- Interact with in-house trading and risk management teams to ensure that the generation gets the best market access possible to maximize their return on the generation investment. Generally, in these situations it is not uncommon for the market bids to be submitted from one of the applications embedded within the GMS.
- Provide data to in-house trading teams to support key activities like settlements, volume, and charge management functions necessary to function in a deregulated wholesale market.
- Perform localized optimization analysis among all the generators available in the fleet to run them at the best possible levels across the various regions that they may exist in—with available market data.
- Support the various plant maintenance functions by providing both operational data as well as market forecasting data, so that maintenance can be performed under the right market conditions if appropriate.

8.1.3 RTO/ISO

From an EMS perspective, the RTO's needs can be defined as that of a super-EMS. The RTO (see Figure 8.2) needs all the functions of a traditional EMS—transmission applications, generation applications, SCADA, and so on. A big difference between the EMS at a transmission operator and that of the RTO is that the RTO does not have the control capabilities of a transmission operator. If the RTO operator wants to execute a supervisory control, then they will need to call the transmission operator and ask them to do it. Depending on the situation and level of emergency, this can be either a request or a command.

In addition to the EMS, a significant focus at the RTO/ISO is also the market functions, which include the life cycle of a market operator—including receiving bids from market participants, validating the bids, clearing the market, sending the instructions back to each of the participants, and finally settling the market. Depending upon the type of market, this activity can either be done at the wholesale level or at both the wholesale and retail levels.

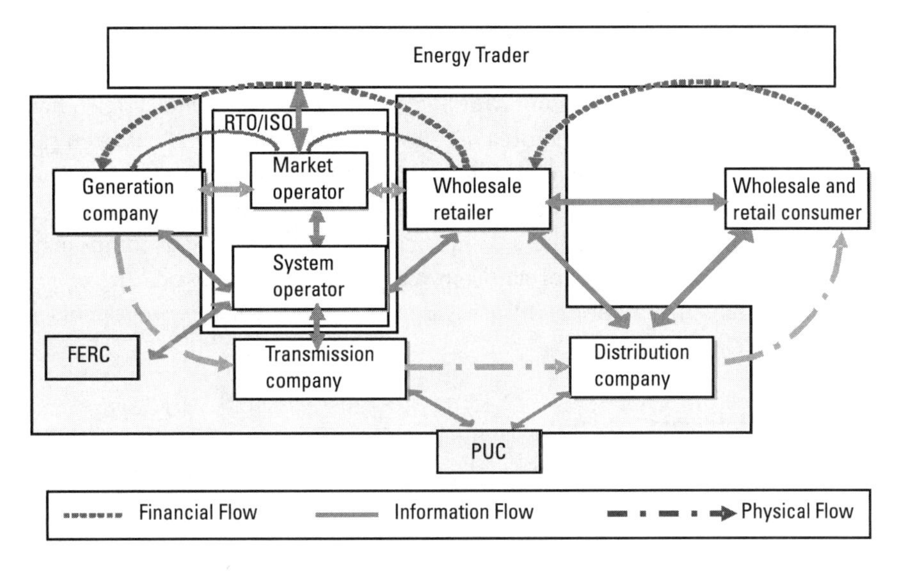

Figure 8.2 RTO and markets interactions with an EMS.

In an RTO/ISO, the EMS and market systems need to be very tightly integrated both in real time as well as in maintaining and managing common models. This is necessary to ensure that the market solutions are compatible with that of the reliability solution because the market clearance for the most part takes system security and congestion constraints into consideration.

8.1.4 RTO/Wholesale Market Participant

The wholesale market participant is somewhat of a unique user of the EMS. Under normal circumstances, this would not be something to even think of—but let us consider the way most wholesale markets function. Market participants bid into the market—from a generator perspective, getting the bid in at the right price is critical. For this they need to know something about system congestion and where it will happen.

For the wholesale participant to know this type of information, it is important for them to also perform some levels of power flow/network analysis. The added constraint here is that the participants cannot have access to the same sets of data or the models as the transmission operator. They need to depend on public models and sources of data and use them to drive market decisions.

Most major market participants have an EMS-like system that is managed based on public sources of data and its outputs are used to drive the market bids. They use this system to predict locations of congestion, and leverage this information to adjust their bid into the market accordingly.

8.2 Key Components of an EMS

The complete EMS system along with SCADA and the front-end systems have been identified as one of the more complex systems deployed. It has been said that the missile defense systems and the space shuttle systems are a few examples of systems that could be considered to be more complex than the EMS.

The typical EMS system is a complex combination of several components that consist of a broad range of hardware, software, databases, and UI.

The following sections will provide insight into these components in more detail.

8.2.1 EMS Hardware

The hardware used in an EMS is varied. However, it can be roughly divided into the following categories:

- Computers servers,
- UI devices;
- Mapboard;
- RTUs and other front-end devices;
- Ancillary devices;
- Communications hardware.

8.2.1.1 Computers servers

The EMS servers are the main computers on which the various application sets are executed. This is also the core location where the database services, UI services, mapboard services, and a host of other system services are executed.

EMS server architectures have changed dramatically over the last 20 years, moving from a monolithic set of computers configured in a dual or quad redundant setup to today's fully distributed architectures designed with advanced redundant fault-tolerant systems capable of being dynamically load-balanced for optimum performance under different operating scenarios. These servers support a combination of Windows, UNIX, LINUX, and other operating systems as desired by the corresponding application set. A typical server setup is presented in Figure 7.5.

A new breed of computer systems is beginning to show up in EMSs: SCADA front-end processors (FEPs). These are systems that have allowed the SCADA systems to become very distributed. The configurations allow these FEPs to be situated anywhere in the network and act as both aggregators of data as well as perform some SCADA processing. The SCADA processing performed by FEPs includes (but is not limited to) conversion from raw to engineering units, error checking, limit checking, and issuing alarms. The significant benefit of this configuration is mainly reduced loading on communications lines leading to a reduced need for communications bandwidth.

8.2.1.2 UI Devices

The UI devices also known as operator's consoles are generally placed on the operator's desk for the purpose of interfacing with the EMS and other systems within the control center. A console is generally a multiheaded (this could be one monitor also—generally this is configurable based on the needs of the user) set of computer monitors controlled by one keyboard and one mouse designed to allow the operator to view multiple systems, views, and applications at the same time.

The operator's console has also transitioned over time from the AYDIN consoles of the old via workstations monitors to the PC-based monitors of varying sizes designed to support the specific operator's needs.

8.2.1.3 Mapboard

Everyone knows the mapboard. Walk into a utility's control center and the first thing that you notice is the huge display of the current state of the utility's operations. The mapboard presents an overview of the system state at a high level. Mapboards have made dramatic transitions over time.

The original mapboards were paper-based (or tiles)—basically a printout of the system one-line (three-line in the case of distribution) connection diagram attached to a wall Some control rooms of today (especially distribution) still have these as their mapboards. Mapboards have transitioned from these paper maps via tile-based (also static) to the newer and extremely dynamic TV/computer monitor-based systems that can be used to project any EMS display or other displays such as weather maps.

There have been numerous discussions on whether the time of the mapboard has come and gone given that rather large displays can be placed right at the operator's console at very little additional cost. Does this mean that the next generation of the control center will no longer have mapboards or they will slowly transition out over time? While the actual answer is unknown, it is clear that in the near- to midterm, they will remain in control centers (see Figure 8.3).

8.2.1.4 RTUs and Other Front-End Devices

RTUs and IEDs are devices that collect data from data acquisition equipment in the field and send them to the main systems (EMS or DMS) over a wired or wireless network. They form the eyes, ears, and the hands of the EMS. They gather data and implement the control signals sent out by the control center operator to perform actions such as closing a circuit breaker, raising/lowering

Figure 8.3 An older mapboard with tiles and static information. (Picture courtesy of Alstom Grid and used with their permission. Alstom Grid retains all copyrights for this image.)

the output of a generating unit, and so on. RTUs come in various configurations and are made by various vendors, each with their own specific protocols. The protocols are used by the SCADA system software to talk to the RTU. The newer ones use a standardized protocol called DNP3.

The key to RTUs and IEDs is that they collect data from local sensing devices that are generally hardwired locally and generally within the substation with few exceptions [5]. The RTUs and IEDs then transmit the information back to the central location through a variety of mechanisms- some wired (powerlines, optical fiber) and some wireless (RF, microwave, or others).

8.2.1.5 Communications Hardware

Communications is a key ingredient of the EMS. Communications are needed from the sensors to the RTUs/IEDs, from RTU/IEDs to the FEPs, and from FEPs to the main system in the control center. Over time, the communications mechanisms used have transitioned from hardwired coaxial cables to radio, microwave, fiber-optic cables, and more recently to cellular/IP. It is not uncommon to find a mixture of all of the above in any modern EMS systems.

Telephones to support human interactions between the control center operators and the field/other utility/nonutility personnel are also a key part of the communications that are part of the EMS. Depending on the level of integration with the EMS, key conversations that are (for the most part) all recorded can be brought up in synch with any specific events in the systems and are generally used for postmortem analysis.

8.2.1.6 Ancillary Devices

In addition to the hardware-based devices that have been described above, there are several ancillary devices in an EMS. Among other things, they provide several basic systems services for the EMS.

- *Printers and plotters.* While the functions of these pieces of hardware are generally self-evident, the presence of plotters in the control center is mainly to support the printing of one-line diagrams.
- *Alarm annunciators (visible as well as audible).* A key application in an EMS is the alarm application. Mainly the SCADA but also many other applications also link up to this application. The key focus of this application is to alert the operator of any anomalies in the system. Many of these alarms are also linked to annunciators, which are generally placed on a wall somewhere in the control center. The intent of these annunciators is to create a visible and/or audible means of alerting the operator of a potential (or existing) problem in the system that needs quick attention.
- *Strip-chart recorders (mechanical as well as PC-based).* There is a predefined set of EMS data for which it is more important to track for their trends versus the actual value. This kind of data is integrated with a strip-chart recorder system (see Figure 8.4). Most control centers will also have a bank of strip-chart recorders (similar to a bank of annunciators as described above) that are checked every now and then by the operators, supervisors, or both. The information in these recorders become even more important when there is a disturbance in the system and the investigators want to see the trend of certain key system variables over time.

8.2.2 EMS Software

The software used in an EMS is also very varied. It can be roughly divided into the following categories:

- Operating system software;
- Application software;
- System and utilities software;
- Network/communications tracking software;
- Process management software.

Figure 8.4 Photograph showing use of strip-chart recorders and other digitally supported devices in a control center. (Picture courtesy of Alstom Grid and used with their permission. Alstom Grid retains all copyrights for this image.)

8.2.2.1 Operating System Software

EMS systems started life in extremely proprietary operating systems also known as mini- and superminicomputers. These were needed because operating systems like DOS, Windows, and others were not yet developed.

The transition from these proprietary operating systems to general-purpose operating systems happened in the early to mid-1970s [6] as these systems moved to the IBM 370 mainframe, Digital Equipment's PDP series of computers or the Control Data Corporation (CDC) computers. At this time, even though EMSs were not on proprietary operating systems, there was little to no ability to interconnect the systems together in a meaningful way.

Over time these systems have moved through UNIX-based operating systems to the modern EMS, which are all based on a diverse set of operating systems architected in such a way that it is more optimized for functionality rather than a monolithic architecture on one operating system.

Operating systems are critical to an EMS because of its real-time nature, more importantly on the SCADA and other front-end systems. To support these real-time functions, EMSs tend to go beyond the normal applications to use specific aspects of system services provided by these operating systems. As a result of this dependence, they are configured to specific versions of specific operating systems and tend to go through major upgrades whenever the operating system needs to be updated.

8.2.2.2 Applications Software

EMS applications run the gamut from SCADA (real-time) to advanced network and generation applications to the more business-focused applications.

- *SCADA applications:* These applications tend to be the truly real-time applications in an EMS and are developed using the more advanced languages of the time, starting originally from assembly language and moving through C/FORTRAN to the more advanced languages of today that include C++ and C#.
- *Advanced network and generation applications:* These applications are some of the workhorses of the EMS and are also the reason SCADA systems were upgraded to be called energy management systems. They are highly algorithmic applications that involve intense mathematical computations using extremely advanced concepts. Some of the key algorithms used here include the Gauss-Siedel and Newton-Raphson methods for solving a nonlinear system of equations, L-U decomposition for storage and manipulation of large sparse matrices, weighted least-square method for state estimation, and regression analysis for predicting and (load) forecasting, among many others.
- *Business-focused applications:* These applications have moved from their traditional EMS-based roots into general relational database-based applications mainly focused on business intelligence, reporting, and so on. In their earlier versions, they were designed and built like other EMS applications and were somewhat limited and simplistic in their capabilities. They were designed in such a way that they could be called on the monitor as a display and then allowed to be printed as a report. Some of the earlier NERC A1/A2 and B1/B2 reports were built like this. The transition to the newer relational database-based applications has allowed them to be extremely flexible in their capability and hence have very strong feature sets. Data from the EMS now gets extracted onto relational or other databases and these applications then function on them.

8.2.2.3 System and Utilities Software

System and utilities software are designed from a system perspective and are capable of being used by other software to either do or check on something.

ALARMS is the single most important application in this set. Most of the EMS applications use the ALARMS application when they need to send an alarm. The architecture of the EMS allows for applications like this to be designed and developed and callable by other applications that will pass the right set of variables into the call. When the alarm shows up in the alarm log (see Figure 8.5), it is generally clear as to which application was responsible for sending the alarm and a description of the problem. Many of the alarms applications adopt intelligent techniques (intelligent alarm processing) to reduce the onslaught of alarms that may overwhelm the operator.

Other examples of this software include configuration manager, console, and mapboard control.

8.2.2.4 Network/Communications Tracking Software

Critical to an EMS functioning is a full and unobstructed access to data from sensors and the ability to send controls to devices in the field and have them complete their actions successfully. For this to happen properly, EMS has a set of software that tracks and manages the availability and condition of the communication devices and channels. When one or more of these devices are not available, they (also) send an alarm that will then result in someone checking out the system for faults.

8.2.2.5 Process Management Software

The EMS is a complex system with applications functioning in a variety of ways:

- Applications like those in the SCADA subsystem execute every 2 to 4 seconds;
- Applications like those in the network subsystem execute every 5 minutes or so unless there is a change in the power system, at which point they stop executing the previous run and start all over again.
- Applications like alarms execute all the time, waiting for some other application to send out an alarm. A typical alarm display is shown in Figure 8.5.
- There are reporting applications that run on a schedule—once a day, at midnight, or once an hour on the hour.

Every EMS has an overarching piece of software system—called the process manager (or something similar)—that manages and controls the execution of all the types of applications.

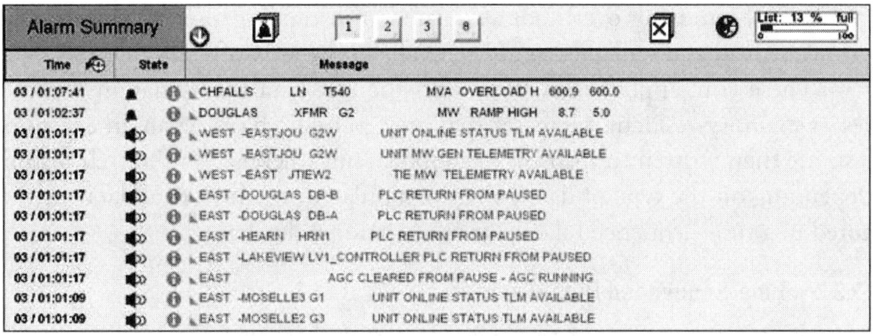

Figure 8.5 A typical alarm display. (Picture courtesy of Alstom Grid and used with their permission. Alstom Grid retains all copyrights for this image.)

8.2.3 EMS Databases

In general, every software application at some level is a database application. EMSs are no different—databases form a very integral part of an EMS. The main strength of an EMS comes from its vast repositories of data and its capability to handle vast quantities of data extremely fast. Database (in an EMS) have come a long way since the days of the original EMSs. They started out from the original file-based databases (many EMSs still have them) to more exotic ones covering the entire range of databases from memory resident to relational.

EMS applications have widely varying needs of their databases that are based on how they function and how quickly they can be used to access data either at the application level or at the display (UI) level. Because of this they can be broadly divided into three main categories: memory-resident databases, time-variant data historian, and relational databases.

8.2.3.1 Memory-Resident Databases

Most of the core EMS applications (SCADA, network, generation, etc.) have an interesting characteristic. They are designed to function in the here and now. The applications are designed to overwrite the previous results every time they complete their execution. They are also designed to manipulate large quantities of data many times in a single execution cycle and in an extremely rapid mode. Most EMSs have some form of a memory resident database to allow this kind of a rapid data manipulation.

Memory resident databases are developed to achieve very low response time and very high throughput for performance-critical systems. They support an architecture to maintain all the data in the main memory and direct data manipulation. All its data is stored and manipulated exactly in the form used by the application, removing overheads associated with caching and translation. As a result, the read and write access are generally at the level of a few microseconds.

These core applications all rely on the fast data manipulation capabilities of memory-resident databases that on a periodic basis or on an exception basis are then written to the disk to support fault tolerance in the architecture. Depending on the type of data being manipulated, the historical data is either stored in a time-sequenced database or a relational database.

8.2.3.2 Time-Sequenced Data Historian

A significant percentage of the data being stored in an EMS is of a time-sequenced nature. For example if we look at the SCADA data acquisition mechanism as explained in the previous mechanism, the application processes the incoming scan and stores it in the memory-resident database. When the next scan comes in, it will overwrite the previous scan. EMSs have a need (for reporting and other postmortem analysis) to store historical data.

If the historical data is of a time-sequenced nature—for example, all critical data from one SCADA scan—it makes a lot of sense to store it in a data historian. Most EMS data historians are designed to store time-sequenced data very efficiently both from a disk space as well as from a retrieval time perspective.

Over time, these time-sequenced data historians have become sophisticated enough to display trending information on EMS displays as well as produce very high-level reports.

8.2.3.3 Relational Databases

Relational databases, once the workhorse of the business world, have been making a slow and steady march into the EMS world. While their mainstay is still the storing of nontime-sequenced data and focusing on corporate and other reports, they have been making inroads into EMS.

Possibly the main inroad has been along with the entry of third-party applications that are being integrated into the EMS. Most third-party applications that get integrated into the EMS come with their own needs for relational databases.

8.2.4 EMS UI

The EMS user interface has changed quite a bit since its origins when EMS came with Aydin (or similar) consoles. These consoles only supported character graphics displays and they needed to be optimized to take up the entire screen size. The configuration was somewhat fixed but these displays were made quite rugged to suit control center environments and their fault-tolerant needs and be capable of driving multiple display units from one keyboard (and a mouse if appropriate). While somewhat uncommon it is still possible to see these consoles in some of the older control centers that are potentially in the process of being upgraded.

The newer UI consoles are mostly either windows-based or Linux-based and provide much of the features we are all used to in our PCs and other workstations. The UI systems are configured in such a way as to enable both EMS and non-EMS displays to be shown on the same workstation monitors

8.3 EMS Application Suites

The EMS (see Figure 8.6) has several application suites integrated into it. As can be seen from earlier sections, it is not a single monolithic set of applications. As EMSs get delivered, it is possible that not all the suites of applications are included. Similarly, other suites of applications (some external to the EMS) may also get added to the EMS depending on the requirements of the client utility.

8.3.1 SCADA

SCADA, as mentioned throughout this book, is the eyes, ears, and arms of EMS. Most external field data from sensors come in through SCADA and similarly most controls sent to operate devices in the field go through SCADA. The SCADA application suite consists of the following main applications:

SCADA front-ends: These are preprocessors for the SCADA data acquisition application whose biggest importance comes from the ability to distribute this application (and associated hardware) around the network. The front-end processor performs most normal data error checking and sends data back to the main server as appropriate. This ability to distribute the application around the network results in significant reductions in network traffic.

The SCADA front-end's other main objective is to interact with the various RTUs and IEDs. This is a reasonably complicated matter because of the number of different protocols that are supported by the various RTUs in the marketplace. While the newer ones are slowly getting standardized into the DNP3 (or DNP3 over IP) protocol, there are several RTUs in the market with several hundreds of legacy protocols that must all still be supported until they get replaced with the newer ones.

Scanner: The data acquisition system feeds into the system through the scanner application. Scanner converts the data from the raw format to its corresponding engineering units and for the first time we actually know that the data coming in from the sensors is actually a voltage, MW, or MVAR reading. Scanner also performs more detailed error checking of

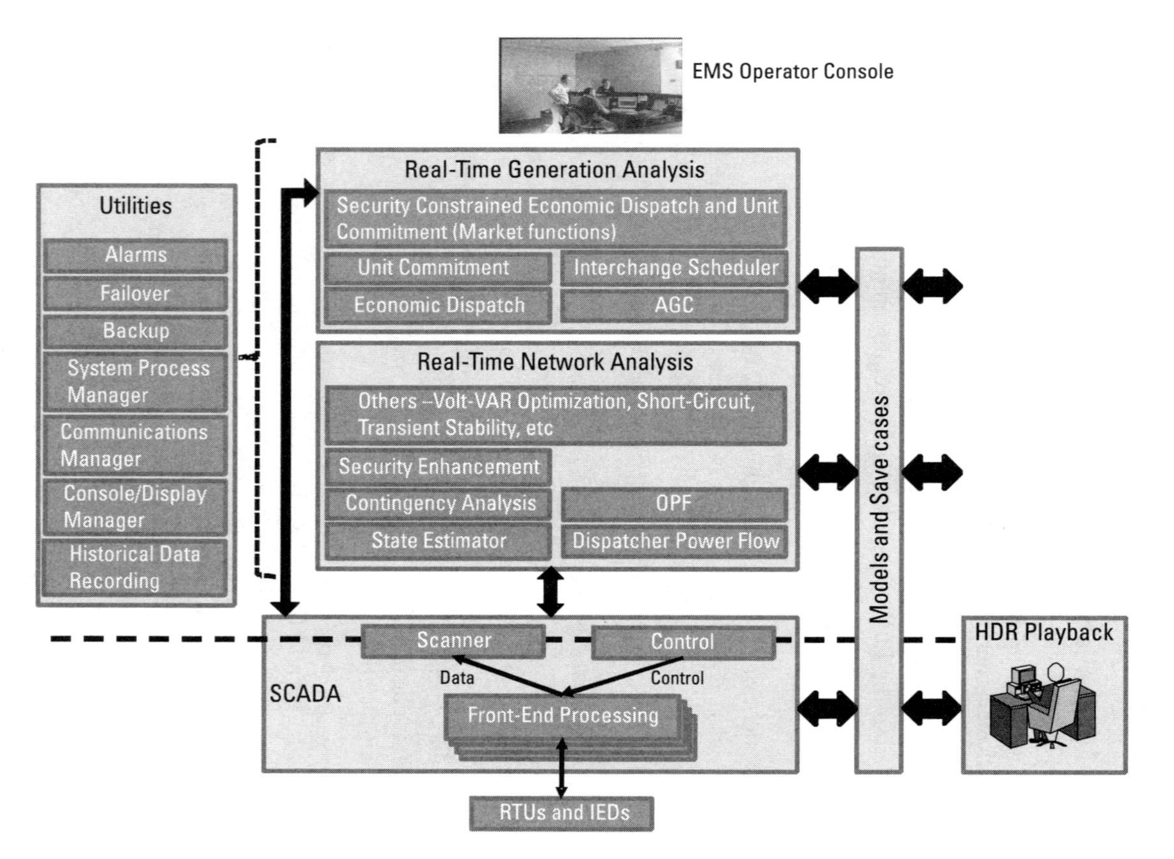

Figure 8.6 EMS application architecture. (© Modern Grid Soltuions.)

the incoming data such as limit checking and some preliminary connectivity checking [7]. When limits get violated, alarms are issued.

At the end of the scanner execution, the data is made available for the various system one-lines and other displays. At the end of this scan, the data is also made available to all other applications in the system.

Supervisory control: All controls into the field go through this application. This includes not just controls manually sent by the operator but also those sent automatically by other programs; for example, generator set-point and pulse controls that are generated by the AGC application. Sending all the controls through one application allows them all to be tracked in a common manner.

A key aspect of the control function is also in its integration with the scanning function where it can send out an alarm if it finds that a specific control sent out into the field has failed to generate the expected result. When that happens, it will also flag the specific control for more investigation and also possibly make it inactive until it is resolved.

Tagging: Whenever maintenance is being performed on a device in the field, the device is first deenergized and then a tag is placed on the device informing all personnel in the field that the specific device is not to be energized, or detailing a specific list of actions on that device that would be prohibited. Similarly, an electronic tag is placed on the device in the EMS as well. The electronic tag will also be accompanied by appropriate notes on the activity that is being performed on the device and it's estimated back-to-service date.

The placement of the tag on the device in SCADA makes the tag visible on all the displays where the device (or device information) is being displayed and operator remote action on that device is prevented.

Managing the operator interface: The SCADA operator interface is an important one in that all key operator actions need to be captured and appropriate response provided. SCADA is specific in this because of the critical nature of the actions being performed from this application suite. A selection of the key operator actions managed under the operator interface include:

- *Removing/restoring key data points from the scanning process.* When this action is performed, that specific (or group of) points will no longer be scanned by the scanning function until it is restored back into the scan again.
- *Inhibiting alarms.* Under some circumstances, the operator may inhibit alarms for some specific set of data points and/or device. This could happen, if for example, a device is under testing and the opera-

tor does not want to be bothered by false alarms taking their attention away from the real problems.

- *Other examples of operator actions:* Changing limits on specific data points or manually overriding the data.

Sequence of events analysis: SCADA scans are dependent upon the time cycle at which the sensors are scanned. This could be somewhere around 2, 4, 6, or 10 seconds and sometimes even more. In between these scans, sometimes the sensed value or the state of a switch (circuit breaker) can change multiple times. This type of rapid changes tend to happen generally during storms and sometimes even during localized problem events.

When this kind of a scenario occurs, there is a need (after the fact) to understand the actual sequence of events that happened so that the exact problem could be identified and fixed.

Historical data recording: SCADA as described earlier is the primary front-line application for the EMS. If SCADA data is available, then the rest of the EMS data sets (for the most part) can be re-created. With the EMS's real-time nature, there is always a need to go back in time and understand a specific event and learn all that happened—either from a training perspective or to identify the core problem and fix it.

SCADA has the ability to store data for historical analysis in a time-sequence manner using a special application and special data sets to store the data. The intent of this storage is to allow (in the future) to run through time in slow motion and truly understand the events that either happened or events that caused a specific series of actions to take place.

8.3.2 Network Apps

The core network applications consisting of the state estimator, power flow, contingency analysis, optimal power flow, and security enhancement all start with a SCADA scan and are intended to take the knowledge of the system state to a level higher than that of the SCADA system and also to override known and unknown deficiencies in SCADA data. They are also intended to analyze the system state and the level of vulnerability of the system to new and potential problems that could happen on the system and prepare the operator (and the system) ahead of time.

Network applications can be run either in real-time mode or in study mode. In study mode, they are designed to provide the operator or an operations engineer with an off-line analysis environment where the same tools as in real-time can be executed but they are run on data sets that have either been stored for analysis ahead of time or it starts with a snapshot of the real-time dataset. The key to this is the need to perform the network analysis in an offline mode and not have a direct impact on the system state at hand.

More information on these applications has been provided in Chapter 7

8.3.3 Generation Apps

The fundamental objective of the generation applications is to provide a mechanism to control the outputs of the generators in such a way as to allow them to provide the necessary supply in an economically viable manner.

The balance of supply and demand is the most important aspect of the electric grid. There are multiple loops in the overall generator control mechanism.

- The innermost loop (using the coal-fired generator as an example) is the one between the boilers, the development of superheated steam that is then fed to the turbine that enables the generator to generate electricity. If more superheated steam is fed into the turbine, more power is generated.
- The next outer loop is the governor loop. The frequency of the power system is reflected in the speed of the generators connected to the power system. The governor is a device connected to the main shaft and controls the amount of steam being fed to the turbine. When the frequency of the system goes above or below the nominal value, the governor reacts by sending less or more steam into the turbine, thereby controlling the output electric power of the generator. Generally, there are several hundred generators in a typical interconnected power grid. Each of the generators will have a governor that will act based on its relative inertia to enable the generator to respond appropriately.

  So far, the generator response discussed has been based on pure physics. If there is a disturbance in the system, the governor loop will act quickly to bring the system to a stable state in the fastest manner possible.
- The next outer loop is the first time economics comes into the picture. This is the AGC loop. In this loop the relative economics of each generator comes into play, thereby redistributing their outputs in proportion to economics instead of their physical characteristics.
- The last outer loop is the markets loop. Here the market offers and bids are considered and the clearance mechanism drives which generator will input more electric energy into the power system. This outcome is completely based on market dynamics and may even have nothing to do with the actual economics of operating a specific generator.

More information on these applications has been provided in Chapter 7.

8.3.4 Dispatching Training Simulator

Modern power systems are becoming increasingly complex to operate. The training departments need more tools to help them in the process of training the operators. In addition, they also need more help in evaluating the training session. This creates an iterative process, whereby the instructors can evaluate the feedback and modify the training program appropriately to cover any holes identified, or identify the weaknesses of each trainee and help them perform better.

A DTS is a simulation of power system behavior and dispatcher user interface (see Figure 8.7). The simulation of the power system provides the operator with a realistic environment in terms of power system behavior. This simulation combined with an exact replica of the EMS applications and user interface allows the dispatcher to practice operating tasks and experience emergency

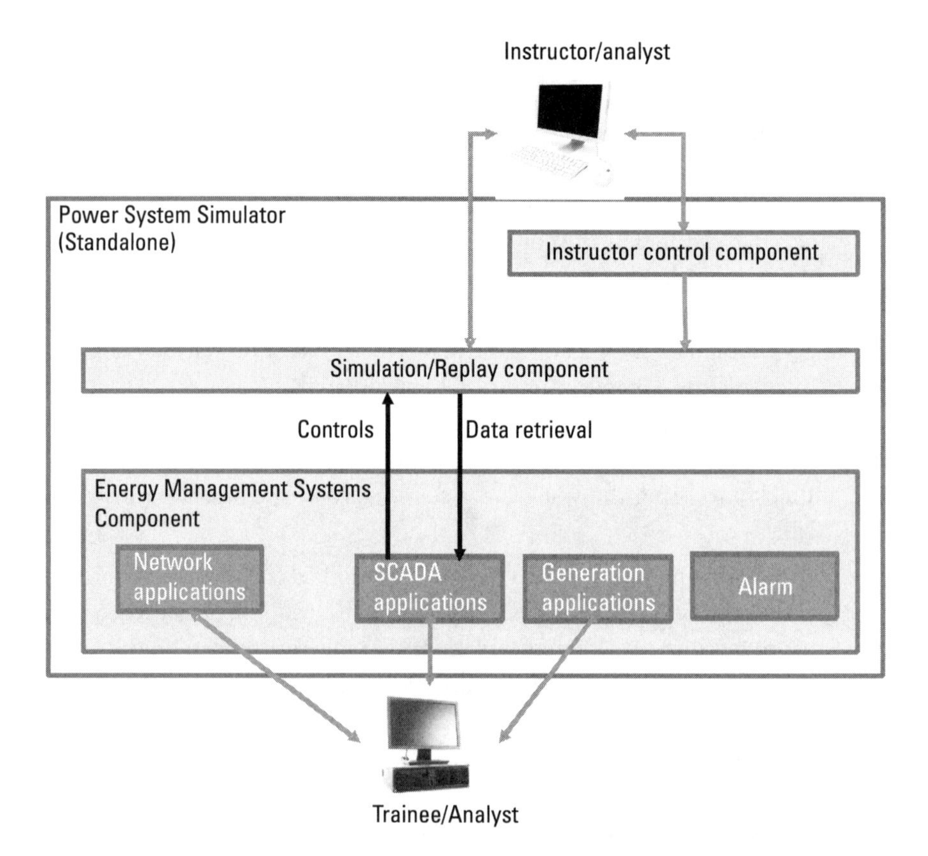

Figure 8.7 High-level view of a dispatcher training simulator. (Picture courtesy of Alstom Grid and used with their permission. Alstom Grid retains all copyrights for this image.)

operating situations. The dispatcher interacts with the computer simulation to practice normal operating activities such as load following, interchange scheduling, supervisory control, voltage control, and transmission dispatch. The simulator is especially advanced in the degree of realism with which it allows the operator to observe and practice procedures for normal and abnormal conditions. Typical abnormal conditions include voltage degradation, line overloads, cascading outages, system islanding, with large frequency deviations, load interruptions, and system restoration.

The DTS also has an instructional system that allows an instructor to create the scenarios on which the trainee (operator) can practice operating the system. The greatest benefit is usually obtained from the interaction of the dispatcher and the simulation. Often the instructor can "assist" the simulation to make it more realistic or interesting. In this interaction, the behavior of the power system becomes very unpredictable and flexible analysis tools are useful for the instructor to modify or analyze the training session.

8.3.5 WAMS

A wide-area monitoring system is a new set of applications tools that are emerging to address not just immediate reliability concerns but also operations issues and long-term system planning as well. WAMS is an enabling technology, and while its contribution to preserving grid integrity in an emergency is clear, it also represents an equally important advancement for the analytic and planning activities that will preserve grid integrity over the long term.

WAMS-based technologies elevate the network applications from performing point-based network analysis to a wide-area analysis of the system. WAMS also has the ability to go beyond to analyze the effects on the system from impending major disasters.

WAMS can have a significant impact on day-to-day operations as well. Having a more precise understanding of the conditions at a specific intertie, for example, would allow an operator to push that connection closer to its operating limit without sacrificing reliability.

8.3.6 Modeling Apps

For EMSs to function properly, they need three main sets of data: (1) the dynamic data that comes from the SCADA scans, (2) the calculated data that is the output of various calculations that are the outputs of the applications, and (3) the system models. Until now, we have mainly discussed the first and the second. In this section we will discuss the third—system models.

For every application in an EMS, we need the static data about how the system is connected so that the algorithms can work on them. For example, for all network applications, there is some basic static information that is needed about the network: (1) the type of component (transmission line, transformer, circuit breaker, etc.), (2) component ID, (3) its location (name of the substation where this component is located), (4) its characteristics (e.g., for a transmission line, its resistance and inductance), and (5) its connectivity (meaning who it is connected to). With this information and the state information as the output of the state estimator (for example), the power flow calculations can be performed.

SCADA models. At its very basic, the sets of data necessary in SCADA models fall into three main groups: substations, RTUs, and communications.

- *Substation and data points:* In this group, the model tracks all the components in a substation and what kinds of data points and controls exist for those components. Information like a specific transformer has x number of sensors for current sensing, voltage sensing, and oil temperature sensing, and if a circuit breaker has y number of controls like open, close, they will all be tracked here. Key characteristics of the sensor like conversion/calibration parameters are all tracked in this group.
- *RTUs:* This group tracks which sensors/controls are coming across which RTU and the specific RTU characteristics that are required to track the data coming across them.
- *Communications:* Communications are a key component of the SCADA model. As a result, tracking their availability becomes equally important. The communications model tracks which RTUs are brought into the SCADA front-end through which communication path.

Network models. The main network model runs in one single group fo-cusking on how the various network components are located—mainly at a substation level. Every component is tracked through the substation they reside in. This works pretty well for most components except the ones like transmission lines, which go from one substation to another.

The models generally supported in the network model include transmission lines, transformers, circuit breakers in all its flavors (circuit breakers, disconnects, etc.), voltage regulators, capacitors, and loads. In general, the load model is considered one of the most important of all—its fidelity will drive much of the fidelity of the calculations in the applications. In general, the level of detail in the model for each component will depend on the various algorithms being supported in the network applications.

While the static version of the network model is collected in a node-circuit breaker version, which means that every component is connected to another through a node and circuit breakers separate the components. The online, or the version of this model used in various software systems, is also called the bus-branch model because the circuit (or any form of a switch) is either closed or open. Based on this, the connectivity model is created.

Generation models. While the generation model is very similar to the network model in its structure, the key focus of the generation model is less on the network/power system attributes and more on the generation economic attributes. Without a market, the economics would be used to perform economic dispatch and confirm that the system is being run in an economic manner.

DTS models. DTS models support the modeling requirements of the DTS specifically. The modeling needs of the DTS are more to support the power system modeling portion of the DTS, while the other modeling needs are met by their respective models.

The primary requirements of the DTS model are focused on the generator prime mover models (e.g., turbines and boilers), protection relay models (over/under current relays, over/under voltage relays, distance protection, etc.).

Source database. Different EMSs look at their modeling applications differently. They either maintain each of the models identified above separately or all of them together in one single location. When they are all located together in one single place, the system is also known by a different name—the source database.

Model validation and verification. Every EMS comes with one or more model validation and verification applications whose primary responsibility is to ensure that the various datasets are in synch with each other and are named appropriately, connected to the right components on each side, and have consistent parametric characteristics.

Endnotes

[1] Definition of the EMS–© Copyright Modern Grid Solutions LLC.

[2] The Inter-Control Center Communications Protocol (ICCP or IEC 60870-6/TASE.2) [1] is being specified by utility organizations throughout the world to provide data exchange over wide area networks (WANs) between utility control centers, utilities, power pools, regional control centers, and Non-Utility Generators. ICCP is also an international standard: International Electrotechnical Commission (IEC) Telecontrol Application Service Element 2 (TASE.2).

[3] A service-oriented architecture (SOA) is an underlying software system that is uniquely designed to allow multiple disparate systems to work together in a many-to-many mode. This underlying software includes a set of predefined business functionalities implemented as software that is reusable by different systems at the same time. Implementing an SOA

architecture forces one to rethink the underlying system design completely because of a planned move away from a one-to-one interaction mode, which is the most common form of implementing these systems.

[4] Distributed control systems (DCSs) are dedicated systems used to control a variety of manufacturing processes like generating stations. They are generally localized to a plant and the controls are distributed in the system with each component sub-system controlled by one or more controllers. The entire system of controllers is connected by networks for communication and monitoring.

[5] Figure 8.3 is a picture of an older mapboard with tiles and static information, Courtesy Alstom Grid. Picture placed here with their permission. All copyrights of this picture remain with Alstom Grid.

[6] In some instances, the sensor may be external to the substation and may generally send the data to a communications aggregator somewhere in the system – which in turn sends the data to the central site.

[7] Figure 8.4 is a picture showing use of strip-chart recorders and other digitally supported devices in a control center, Courtesy Alstom Grid. Picture placed here with their permission. All copyrights of this picture remain with Alstom Grid.

[8] Glenn Stagg has been widely credited with being the pioneer who made the first move from special purpose computers to general purpose computers.

[9] Figure 8.5 is a picture of a typical Alarm Display, Courtesy Alstom Grid. Picture placed here with their permission. All copyrights of this picture remain with Alstom Grid.

[10] The connectivity checking is another form of error checking. For example, if a circuit breaker is open, then under normal circumstances the flow of power through a line should be zero. If it is not, then the Scanner can identify that one of the two measurements is suspect.

[11] Figure 8.6 shows a high-level view of a Dispatcher Training Simulator, Courtesy Alstom Grid. Picture placed here with their permission. All copyrights of this picture remain with Alstom Grid

9

Outage Management System

The electrification infrastructure has been hailed as the single most important achievement of the 20th century and that mandate has been provided to utilities to carry it out. The most basic mandate of the utilities is to provide safe, reliable, and high-quality power to its customers. It is indispensable to factories, commercial establishments, homes, and even most recreational facilities. Lack of electricity causes inconvenience and economic loss due to reduced industrial production or in preventing the conducting of commercial business.

The definition of an outage [1] is: a quantity or portion of something lacking after delivery or storage or a temporary suspension of operation, especially of electric power.

As our definition indicates, an outage can be anything from a sustained lack of electric power to temporary loss of power and even to flickering of power (which is more of a lack of power quality). When any of these happen, it disrupts life. When an outage happens, in essence, the utility is not able to deliver on its mandate.

Outages happen due to a variety of reasons, including:

- The circuits in a home get overloaded, causing one or more of the circuit breakers in the house to trip, or the main circuit breaker trips because of overloading in the circuits inside the home. When this happens, either part of the home or the complete home loses supply.
- A line gets overloaded and either burns down due to overheating or the overload leads to a relay or circuit breaker tripping, which in turns leads to one or more circuits being disconnected from supply.

- A line gets overloaded and the overheating leads to the line sagging so much that it may touch a tree or something else, thereby short-circuiting it to the ground and leading to relays tripping and disconnecting parts of the system from the supply.
- An electric pole or transmission tower gets knocked down either due to a storm or due to a car hitting the pole, leading to lines getting disconnected or falling on the ground and short-circuiting.
- A storm blows over an area and several trees or tree limbs fall on electric lines, thereby severing them and resulting in circuits being disconnecting from supply.
- A combination of animals and faulty equipment/human error has also been known to cause outages. Rats or other animals can get into substations and short out the busbars, leading to an outage in a substation which in turn results in an outage for several people who are fed from that substation.

While the list provided above is only partial, each of these incidents (and possibly others) can cause outages ranging from one home to large areas of outage within and beyond a single utility's jurisdiction.

9.1 Types of Outages

There are two main types of outages: transmission and distribution. Their behaviors and impact are vastly different, as detailed below.

9.1.1 Transmission Outages

Transmission systems as we have identified earlier are generally networked in nature. This means that most times, when one component fails either due to a fault or a natural disaster, the system will still stay stable and continue operating as if nothing happened. However, under some severe circumstances, a single outage can cause other lines to be overloaded and then trip, leading to a cascading situation. When this happens, a much larger portion of the grid gets disconnected from supply and every customer (residential, industrial, and commercial) in that area loses power.

The networked nature of transmission systems also can lead to situations where the cascading outages can escalate very rapidly, leading to large segments of the population being out of power very quickly. The 2003 outage in the Northeast United States is one such example, where in a period of 4 hours a large portion of the Northeast part of North America was outaged and about 55 million people lost power across the United States and Canada.

Another major characteristic of transmission outages is the fact that these systems tend to be well covered by sensors and as a result the utility will know if and where the power is out. The system operator also has several tools and the benefit from experience working through restoration drills over several years. Regardless of the type and extent of the outage, these restoration drills allow the operator with strong knowledge of system behavior and different mechanisms to start the restoration process and bring systems back quickly. In the Northeast blackout, while some power was restored by 11 p.m., many did not get power back until 8 a.m. the next day.

9.1.2 Distribution Outages

Unlike transmission systems, mostly distribution systems are radial in nature. As a result, when something trips, most customers downstream from the location of the trip will lose power. Distribution outages also are rarely of a cascading nature given that they lose load that is not picked up by other feeders in a networked system.

Most distribution outages are due to storm or severe heat conditions in which large swathes of a utility's jurisdiction loses power mainly due to downed power lines. Distribution systems also by their very nature tend to be spread out, which makes it difficult to restore because the storms cause several outages and each one needs to be repaired individually with a lot of attention. When a storms comes in, for utilities, it is generally "all hands on deck" with every employee taking on a specific storm response role. For large distribution outages, it is also fairly common for crews from different utilities to come in and help with the restoration process.

A unique aspect of distribution outages is the nested outage, which, as the name implies is an outage within an outage. Nested outages happen because there are actual multiple outages in a neighborhood or location and the one with the largest impact is fixed first. When this outage is fixed, it is not immediately known if there were other outages downstream of this outage until the feeder is energized and someone actually drives through the neighborhood to check if all the homes are energized. It is expected that the implementation of AMI systems may help with solving the problem of nested outages because most AMI meters can be pinged to check if they are live or not.

Figure 9.1 presents the output of an outage analysis superimposed on an OMS outage map. The shaded areas point to locations in the distribution system that have lost power and based on that the system operator identifies the possible location for a fault and sends the trouble-men to check it out.

Figure 9.1 Distribution-level outage analysis superimposed on an OMS map. (Picture courtesy of Alstom Grid and used with their permission. Alstom Grid retains all copyrights for this image.)

9.2 Origins of the OMS

Outages have been in existence since the dawn of the electric utility. To track outages, some form of an OMS has also been in existence since then.

The origins of an outage management system started with the recording of trouble-calls, leading to an earlier name—trouble-call management system. An early version of the trouble-call process was basically to record incoming calls with customers or others reporting a problem in the system. The problem could be of any form, obviously with an outage being one of the more common items reported. In those days and in most places even now, the only way for a utility to know about an outage was if someone called in and informed the utility that they were out of power at their home or premise. The utility collected the information from the outage calls, and from the pattern of calls received and their locations, they determined the likely location and cause of the outage. A crew was then sent to the location of the outage to investigate further and begin repairs.

9.2.1 The Paper Age

When calls were received at a utility's customer service center, they were recorded by hand on a "trouble ticket" or they were entered in a computer and then printed. These tickets were then sent in some form to the service center where the outages were being managed for that location (very often called service districts). At that location, these tickets were manually sorted by the circuit on which the customers were connected and then placed in a group for further action.

In the service centers, experienced analyzers would then look at the tickets and their locations and try to identify the root location of the outage. Printed copies of the electrical maps of the distribution grid would be used to assist them in following the outage locations and identifying the root cause location. A trouble-man (or T-Man) would be sent to the location to pinpoint the exact source of the outage. Sometimes they would start from the location identified in the service center and then drive along the line to locate the real outage. They would then make an assessment of the damage and report back to the service center, where a service work order would be created. If the damage was simple, the T-Man would be able to fix it—if not, a standard bucket truck with the right equipment and supplies would be requested.

This extremely manual set of activities worked well when the volume of calls was light and the number of outages was small. However, during storms when a large number of outages happen, the number of calls goes severely up and this process becomes difficult to sustain. At this time, the utility will also need to prioritize which calls get addressed first and so on, leading to an increased potential for error, lost (or misfiled) trouble tickets, and so on.

Manual paper-based processes also made the collection of outage metrics very difficult because (1) very often after a storm most of the regular employees went back to their regular jobs, and (2) much of the effort at collecting the outaged customer data was difficult to calculate and ended up mostly in guesswork.

9.2.2 The Move to an Outage Management System

As utilities got more customers, there were more outages, there was a greater expectation of the need to resolve them quickly in a cost-effective manner. Enter the beginnings of the modern outage management system.

At first, utilities were just automating the manual process described above. This resulted in the early trouble-call management systems, which started with manual entering of names and locations of outaged customers into a database with all customer data and their locations. As these grew in size and complexity, two major sets of information were added:

1. An underling connectivity model of the distribution system with as much detail as possible all the way to the customer's premise. This was generally stored in the form of a feeder tree, which was a static description of each feeder, and the position of each protective device (e.g., fuse or recloser) in the feeder hierarchy and customers were assigned to specific circuits and upstream protective devices.
2. Simplified algorithms that took advantage of the connectivity model and provided the service analyzers with quicker feedback on the probable location of the root cause fault.

Over time, these systems got more and more sophisticated, leading to the addition of graphical user interfaces with a graphic form of the connectivity model on the screen with outages actually placed on them. Automation was also getting added to these systems to allow for the prioritization of the outages based on key business rules such as number of customers in a single outage or criticality of a specific outage (e.g., for a hospital), and so on. The next step resulted in the interaction and creation of work orders so that either trouble-men or other field crews could be directly dispatched and the resultant restoration of the outage logged against the work order and the subsequent customers.

As the level of automation increased, the outcome from these systems also became more effective and the reports more accurate.

9.3 Architecture of an OMS

The OMS is one of the few systems in this book that has registered users from outside the control center environment. This is more of a business application suite than an operational application suite. The set of utility personnel who tend to have access to this suite of applications include system operators, emergency service dispatch center personnel, trouble-men, utility mid-to-senior-level management and executives, and in the event of a storm, just about most utility personnel. As a result, in a typical utility, it is not uncommon to have several hundreds of users on this system, some inside the control center and many outside. This also means that the OMS as a system is accessed by several people whose access may need to be tightly handled and managed to ensure that they do not cause any inadvertent issues with either the data entry or data interpretation. In addition, the business aspects of this system also requires some extensive controls and audit tracking—basic requirements of any business application.

Figure 9.2 provides a very good insight into the details of how the OMS is set up within a utility.

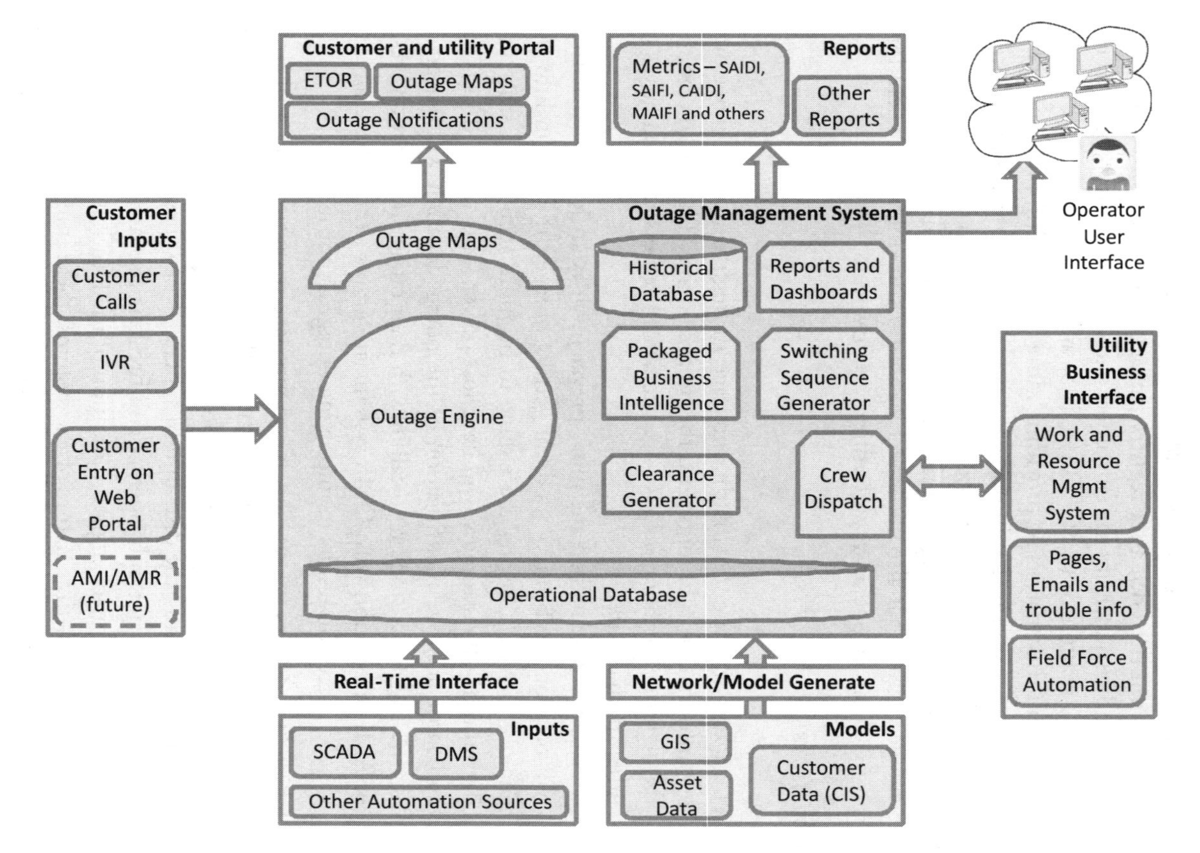

Figure 9.2 OMS architecture. (© Modern Grid Solutions.)

Outage Engine

The engine's basic functionality is to take the input calls and other sources of outages to pinpoint possible locations of the actual outage. During storm conditions, this can be tall order because there could be several thousands of calls coming in reporting several hundreds of outage locations. The engine uses the core database from the GIS, asset registry, and customer locations to perform these actions and also identify the customers and customer counts on who has lost power. Information from SCADA, DMS, and other automation sources can also augment this analysis by providing more concrete information regarding open devices or overloaded information in the field. The GIS information including maps and connectivity also allows the engine to actually place the customers and source of outages on the outage maps, which are used not only by the outage analysts but also on large TV monitors in emergency operations centers for viewing by operators and utility executives.

Key Interfaces

- *Customer interface:* This is the interface through which customer outage information is brought into the outage engine. This could come in through one or more of IVR [2] systems call centers representatives or other mechanisms.

  As AMI systems become more prevalent, it is anticipated that they will become the primary mode of getting outage information into the engine, thereby allowing the engine to be more accurate in truly understanding all customers who have lost power. This will lead to faster and more accurate identification of the outage location without actually waiting for customers to call in with their outage information. Tests of AMI integration with OMS systems at various utilities have already yielded successful results in improving outage response.
- *Real-time data interface:* This is the interface to some of the real-time systems like SCADA, DMS, or other real-time automation (substation automation, distribution automation) subsystems that are also responsible for protecting the distribution system. When these systems trigger on a fault, they not only open circuit breakers that are known ahead of time but also provide a good indication of the location of the fault. Because of the real-time nature of the systems, these interfaces tend to be very specific to the system they are interfacing with and most often these interfaces also tend to be one-way into the OMS. Any return data movement out of the OMS to SCADA and so on tends to be done manually.

- *Network/model generation:* One of the key foundations to an effectively functioning OMS system is access to an accurate network connectivity model and associated renderings (most often) from a GIS system and list of assets and their locations from an asset registry. The OMS starts off with these two sets of models and superimposes various kinds of information on top of it—outages, temporary model changes in the field, system tags, outage statuses, crew locations, and so on. All of this also gets updated on the renderings so that all users can get a visual map of the actual outage situation at any time.
- *Utility business interface:* Outages are such an important and sensitive issue for utility executives that much has been done in terms of automating several of the interfaces to utility business systems as well as providing information to utility execs and managers. The most important of these interfaces are the ones to the work and resource management system. In this system, the outage analyst can take outages as they are prioritized and use this interface to direct work to field crews. Using the field force automation interface, the outage analyst can send out key information and stay in constant touch with the trouble-men as well as the field crew as the outage is being fixed and the circuit reenergized.

  Utility managers and executives also like to stay in the loop on outages and many tend to be on e-mail/pager alert as different outages are identified and the number of customers estimated. This is an important part of utility internal communications so that (if necessary) the executives can jump in and assist with reassigning resources depending upon the nature of the emergency.

Customer Portal

Many OMS systems are being paired up with specialized customer portals that could nowadays include a combination of both Web-based information dissemination and even push of outage notifications and other information to a smartphone if so desired by the customer. From the portal, which is fed mainly by the OMS, the customer can either get general outage information like how extensive a specific storm-based outage is. They can also get customer-specific outages upon some form of a log-in where they can get better estimations of ETOR.

Report

Given the importance of outages, it is normal to expect that several agencies are looking for outage metrics and other reporting requirements from an OMS.

Beyond the normal need to track the foundational metrics like SAIDI, CAIDI, and SAIFI, as mandated by NERC, newer ones like MAIFI are coming up and becoming more and more important in today's utility environment.

Beyond the needs of regulators, utility management also looks for different pieces of information that don't just drive an understanding of the performance of the utility under a major outage, but more important, is also used to drive capital decisions of equipment repair/replacement.

As a result of these reporting requirements, OMS systems tend to have within itself, a fairly complex reporting engine that not only has canned reports but also allows the users to create ad hoc reports for specific information mining.

Operator User Interface

This in fact covers all user interfaces for OMS systems.

OMS user interfaces have evolved dramatically over time. They started out with systems that utilized a tabular user interface that tracked it by the outage and recorded the extent of the outage in each row. However, these mechanisms made it difficult for the user to visualize the outage location. It was also difficult to reflect the effect of any feeder reconfiguration that may have been performed, especially if the reconfiguration was only temporary. Over time, these systems gave way to a spatial representation that was able to bring a graphical approach to outage management.

The newer OMS systems are based on the GIS maps and the underlying spatial data structures. As outage calls come in, they are mapped on to the feeders directly and show up on the map. The maps are both graphical and geospatial in nature—an aspect that is necessary to allow not only the appropriate grouping of outages by feeders but also to provide for the right routing and scheduling of trouble-men or field crew and enable maximum efficiency. In some, the right interfaces with work and resource management systems also allow for mapping crew locations right on the same map as the outages so that the operators know which crew is working on which outage and also who is the closest to the next outages.

In addition, integrating the outage maps to industry standard tools like Google Maps is being tested by several vendors and could become standard offerings in the next generation of OMS systems.

9.4 The Business of Managing Outages

Outages are important business for a utility. Consumers do not like outages [3] and want them to be gone. They speak loud and clear through their state regulators who tend to penalize their utilities during rate cases when they feel that a utility has done a particularly bad job at responding to outages. Lastly, the public perception of a utility's brand is so heavily linked to their outage response, that recently they have been punished even more severely by their shareholders when their outage response (or the lack of appropriate response) reaches the news outlets and becomes a major point of discussion. An example of this could be seen in the resignation of the president and chief operating officer of Connecticut Light & Power, the state's largest utility, after coming under fire for how the company handled widespread power outages following a major snowstorm in 2011.

> People don't like power outages. What may be surprising is the number of Americans that are not willing to tolerate disruption even for some handy compensation. More than one-quarter of the American public believes they should never experience an electric power outage, unless there is an extreme weather event [3].

Given its most often direct link to rate case outcomes and the fact that utilities are fundamentally asset management companies and live and die by their ability to recover on their investments (utilities prefer to do a good job on their outage response), they really do like to restore outages as soon as possible. They do this by several efficiency measures such as:

- *Position crews in key locations ahead of time.* As the path of the storm is identified, utilities are beginning to position key crews (mainly trouble-men to start with) in locations spread out in their jurisdiction so that they can start the restoration work as soon as possible. Some utilities are also having mobile service and dispatch centers that can be dispatched to other locations to increase the number of people who can dispatch. This becomes even more important during large storms like hurricanes where they may also need to coordinate with crews from other utilities.
- *Dispatch crews and trouble-men more efficiently.* Sending the right crew the first time with the right equipment and parts is an important aspect of changing the paradigm. This can be done by providing the trouble-men with OMS terminals and they can perform a full diagnostic and assessment of the problem and the parts needed. This requires either the field crew who go there after the trouble-men to first come back to the service center to get everything they need or for them to go straight to the problem location and have the parts and tools sent to them separately. Either of these mechanisms can work depending on where they are all located.

- *Reduce okay on arrivals (OKA).* OKA is an important aspect of outage management because truck rolls cost time and money that could be better spent on solving actual problems. Fully confirming that there is a real problem is critical to sending the crew out there. Admittedly, this is a difficult problem to solve, though AMI should help in alleviating this to a good extent.
- *Improve ETOR timelines and accuracy.* This means having a structured process for estimating and updating the ETOR information. From a customer service perspective, people tend to place a lot of emphasis on getting their power back within the time frame given by a utility—and when they do not get the power back on time and if it happens very often, they lose faith in the utility's ability to predict when their power comes back on.

9.5 Impact of Smart Meter on the OMS

Smart meters have the potential to make some of the most dramatic improvements in the arena of outage management. For the first time, utilities will have something at the home's location that knows when there is power to the home and when there is not. These meters also have the ability to detect if the problem is on the premise side of the meter or on the grid side.

Smart meters have some key characteristics [4] that allow them to provide the features described above and many more, including:

- *Last gasp:* Smart meters have the capability to send a "last gasp" signal to the metering head-end system just before they lose power. The meter head-end system can pass this signal on to the OMS which in turn can interpret the premise as being without power. This information is used by the OMS in the same way that it would process a customer call with information of loss of power. This characteristic allows the utility to know about a customer loss of power even before either the customer knows about it or calls the utility about it. This will allow the utility to inform the customer before they call in, thereby reducing call center call volumes. Getting this kind of information on all the outages in a location allows the OMS to better understand the extent of the outage and hence better diagnose the probable location of the originating fault and send the right crew to the right location. These characteristics provide for reduced and possibly more accurate ETOR calculations.
- *Ping for being powered:* Smart meters also allow for the meter head-end operator to ping them to see if they are still powered. When the cus-

tomer calls the customer service representative about an outage, they can ping the meter to check if the problem is on the grid side of the meter or on the home side. If it is the latter, then the customer may need to fix the problem by themselves.

Nested outages are a phenomenon that occurs during widespread storm damage. There could be multiple outages and some could be nested—meaning that when there are multiple outages, sometimes the outermost outage could be identified but there could be some inner outage locations as well. As soon as the crew has fixed the outage problem, the operator can ping all the meters to see if they all come back on—and if some of them are still out of power, then it means a nested outage has been identified. Now, instead of the crew finishing one outage and going back to the service center only to find that there was another outage at or near the previous service location, they could move on to the new location and work on fixing the problem. This characteristic allows for overall improvements in SAIDI/CAIDI numbers and of course the obvious customer satisfaction improvements.

These smart meter impacts on the OMS are affected by the types of communications methods used between the smart meters and the meter head-end. If it is wireless, and a part of the wireless system goes down, then it does become difficult for the smart meters to either send out a last gasp or respond to a ping. When something like this happens, then we are back to the days of OMS in which we were dependent on the ability of the customer to call in and tell the utility about the outage, and also possibly the need to call them back to tell them that the power is back on.

9.6 Future of OMS?

The future of the OMS system is a little cloudy. Much of it depends on the success of the DMS (next chapter) and how it is taken up in the marketplace. The OMS and DMS have a lot of features in common as will be explained in the next chapter. These common features are:

- *Reduce manual work for switching, logging, and reporting.* Keeping the full state of the system in one place is important because it simplifies the maintenance process of keeping the two databases in synch. There are several items of data that need to go back and forth between the two systems—device statuses, SCADA maintenance tags, relay/fuse trips, and so on.

- *Tracking all deenergized segments of the grid in one place.* Currently, planned outages and those observed by SCADA are fist maintained in the DMS and then passed on to the OMS. Similarly, the ones not observed through SCADA but called in are recorded in OMS first and need to be either manually or automatically passed on to the DMS.
- *The need to generate switching sequences for either deenergizing a component or for reenergizing it.* It does not make sense to develop switching sequences for unplanned outages in an OMS and for planned outages in a DMS. Developing all switching sequences in one location is optimum because it would then take advantage of the full state of the system that is both planned and unplanned. In the long run, keeping these two systems in synch would require a tremendous amount of effort because the two systems are updated differently and the control requirements of the two are also different. OMS is fundamentally a business system (it does not perform any controls on the system) whereas the DMS is fundamentally an operations tool and performs controls on the system via a SCADA-like interface. However, they both have many common functions and track a lot of common data, and as a result, some utilities are taking the step of using combination systems—a step that even several vendors are also taking.

Much of the information on DMS systems is described in the next chapter.

End Notes

[1] http://www.answers.com/topic/outage.

[2] Interactive voice response system (IVR) is a telephone-based system designed to improve communications and assist customers with their needs. An IVR system is a software application that enables users to interact with a database through the keypad of a phone or voice commands. This allows services to be made available to customers 24 hours a day. IVRs can route calls to in-house agents or transfer calls to an outside extension, in this case to the outage engine.

[3] "Americans say: You Can't Pay Me to Tolerate a Power Outage," http://www.fierceenergy.com/story/americans-say-you-cant-pay-me-tolerate-power-outage/2012-05-11?utm_medium=nl&utm_source=internal#ixzz1uZvas9ES.

[4] Not all smart meter installations will have these characteristics. If you consider the evolution of smart meters all the way from AMR to AMI to today, they will have a variety of capabilities. However, these characteristics and others are expected to be the wave of the future.

10

Distribution Management Systems

Distribution management for the longest period of time basically consisted of a basic SCADA system supported by paper maps generally operating out of one of the back rooms of a utility's service centers. The SCADA system was generally piggybacked out of the transmission SCADA and one or more consoles were provided for the distribution operator to monitor and operate the parts of the distribution system that were covered by the SCADA system.

Unlike the situation with the EMS where there was a clear and specific requirement for the system borne out of the (now famous) New York blackouts, there wasn't a specific industry event that specified a need for a DMS system. The need for a DMS came more from the ongoing increase in sophistication within the distribution system with more and more sensors being added, and supported by customer need for a more reliable distribution system that required an increased ability to control the system closer to the problem area. The arrival of the smart grid accelerated this effort.

In the mid-2000s, the distribution system started getting more attention. The inability to restore outages in a faster manner coupled with an aging workforce alarmed several of the distribution utilities. Their ongoing poor performance metrics in SAIDI/CAIDI/SAIFI have made distribution operation a business platform with rising costs and unsatisfied customers complaining to their regulators. The set of industry drivers are identified next.

Increased Competition and Focus on Regulation

Utilities are facing significant external pressures that are driving a business model change for the industry from a customer perspective. States have been increasingly active in promoting competition and market restructuring since the approval of the Energy Policy Act of 1992. These days, competition is primarily driven by the threat of municipalization [1].

Additionally, advocacy and industry watchdog groups have been highlighting the inadequate service provided by electric utilities, and state public utility commissions react by imposing increasingly demanding regulations as an alternative means to increase customer satisfaction. This has escalated the need for utilities to improve performance and enhance services in order to strengthen existing customer loyalty and attract new customers. Several consumer groups are also recommending improvements to service reliability, stronger enforcement of rules, better consumer protections, and mandating public reporting processes.

Load Growth and Aging Infrastructure

Utilities are witnessing rapid expansion of power distribution systems in size and complexity coupled with a strict limitation of investment resources that severely limits the ability to improve the aging infrastructure. This is coming from having more customers in a service area and more load per customer [2]. Historically, the primary approach a utility would employ to mitigate the effect of load growth and aging infrastructure would be to spend significant dollars to upgrade the electric system. This is no longer possible because of increased pressures on infrastructure spend.

Data Overload

Many utilities have separate user interfaces for operating SCADA devices, managing outages, interacting with field resources, entering outage information for customer communications, and logging and writing switching orders. In storm situations, operators find themselves managing multiple outages, toggling back and forth, trying to keep track of it all. Many times operators find themselves duplicating work effort and entering in the same information multiple times. A distribution system operator is simply overloaded with data. The data comes from different user interfaces and he or she does not have the time to convert the data into useful information that can be leveraged to make decisions. Without the time or tools to make sense of all the data, much of it is wasted, leading to less than optimal operation of the distribution system.

Aging Workforce

The aging workforce has and will continue to be a significant issue in the utility industry. In the next 10 years, the utility industry expects a significant reduction in its workforce as the average age of a utility worker is greater than 50 years old. For some utilities, the number of expected retirements is over 40 percent of its current staff, including supervisory personnel. This departure of key personnel represents a loss of knowledge capital, experience, and capability and results in a huge risk to effectively and efficiently operate the distribution network. The key factor that is keeping aging workforce as a critical driver is the state of the economy, which is forcing many people to stay in their jobs longer.

Within distribution system operations, the long lead time required to onboard a new hire makes this even more pressing. Depending on the utility and the jurisdiction, the lead-time can range from 1 to 2 years to be capable to operate the network. Developing the proficiency and knowledge of the "traps" within the jurisdiction takes an even longer time. All of this intensifies the criticality of the risk to the utility.

Desire to Reduce Outage Duration and Frequency

For a utility, an outage represents a failure to deliver on their basic mandate of delivering reliable power to their customers. As a result, they do not like them and will go to great lengths to either reduce outage duration and/or frequency.

With each outage their customer base is less satisfied with the electricity service and there have been many cases of cities breaking away from their utility. In addition, SAIDI and SAIFI costs are impacting revenues as utilities are paying millions in PUC penalties or their rate-cases are being refused/ reduced.

However, outages are a fact of nature and are in fact most often caused by acts of nature [3]. The way a utility reacts to an outage either through communicating with its customers and/or mobilizing the crews to get to the outages quickly and efficiently says a lot about the utility's interests in reacting to its mandate. Utilities that tend to get hit by large hurricanes a lot tend to get better at responding to them by planning their crew locations around their service territory based on the anticipated destruction leading to loss of power.

Need for Increased Visibility

Visibility into the status of the power system network can have a significant impact on a utility's ability to either support planned work and/or unplanned situations like storm restoration.

The manner in which a utility responds during an emergency is critical not only for safety but also for customer satisfaction. Paper wall maps of the past used to make it difficult to isolate where outages occurred, and know the state of equipment and location of field crews. After many phone calls and sectionalizing circuits, more knowledge about the outage is gained but unfortunately only the control center would have the information because efficient dissemination of data could not be done electronically.

Informing the emergency centers are generally done using phones and as a result response is not quick enough. Without an efficient real-time system, customers are also not updated on the situation, causing more angst amongst them.

Reduction of Risk from Disasters

In the current paper wall map environment, there is no backup map in case of a regional disaster. With paper maps, colored pushpins are inserted into a map of the network to indicate open or closed switches and crew locations. Tagging is done with markings on a paper hooked onto the location to indicate a location where a crew is dispatched.

There have been cases where pins and tags fell off the paper map. In this scenario there is no other backup map to indicate the "as-operated" state of the network. Control centers will spend countless hours calling crews and analyzing switching plans to recover from a disaster situation. If the disaster is more extreme and knocks out a control center completely, there is no way to transfer control to another control center. Cross-jurisdictional operations is more complicated.

10.1 Introduction to the DMS

A distribution management system (see Figure 10.1) is a system of computer-aided tools used by operators of electric distribution networks to monitor, control, and optimize the performance of the distribution system [4].

Currently, in many utilities, the system operator function is fairly manual in nature with several control centers still using paper maps for distribution. Key data such as relay settings are still not fully integrated into the major systems and these are critical to support switching activities. Outage management for the system operator still consists of manual data entry (after receiving a call from the customer) versus proactive identification of outages and their locations. The system operator skill set still for the most part is the journeyman skill set that emphasizes a more physical understanding of the grid instead of the power system knowledge with focus on information management.

In order to face these drivers, utilities are implementing DMS as a tool and are becoming front runners in distribution operations. DMS provides greater monitoring and control over the distribution grid in addition to enhancing efficiencies in the operator's daily tasks, which results in quicker location of faults and lower duration of outages.

These key drivers analyzed have created a situation that results in a call to action to upgrade distribution operations and bring it into the 21st century to take advantage of modern applications and better communications mechanisms.

DMS is answering this call to action!

A distribution management system was developed to provide a framework in which utilities can more easily support and maintain the distribution grid through the use of modern technology. Similar to an EMS, which started with

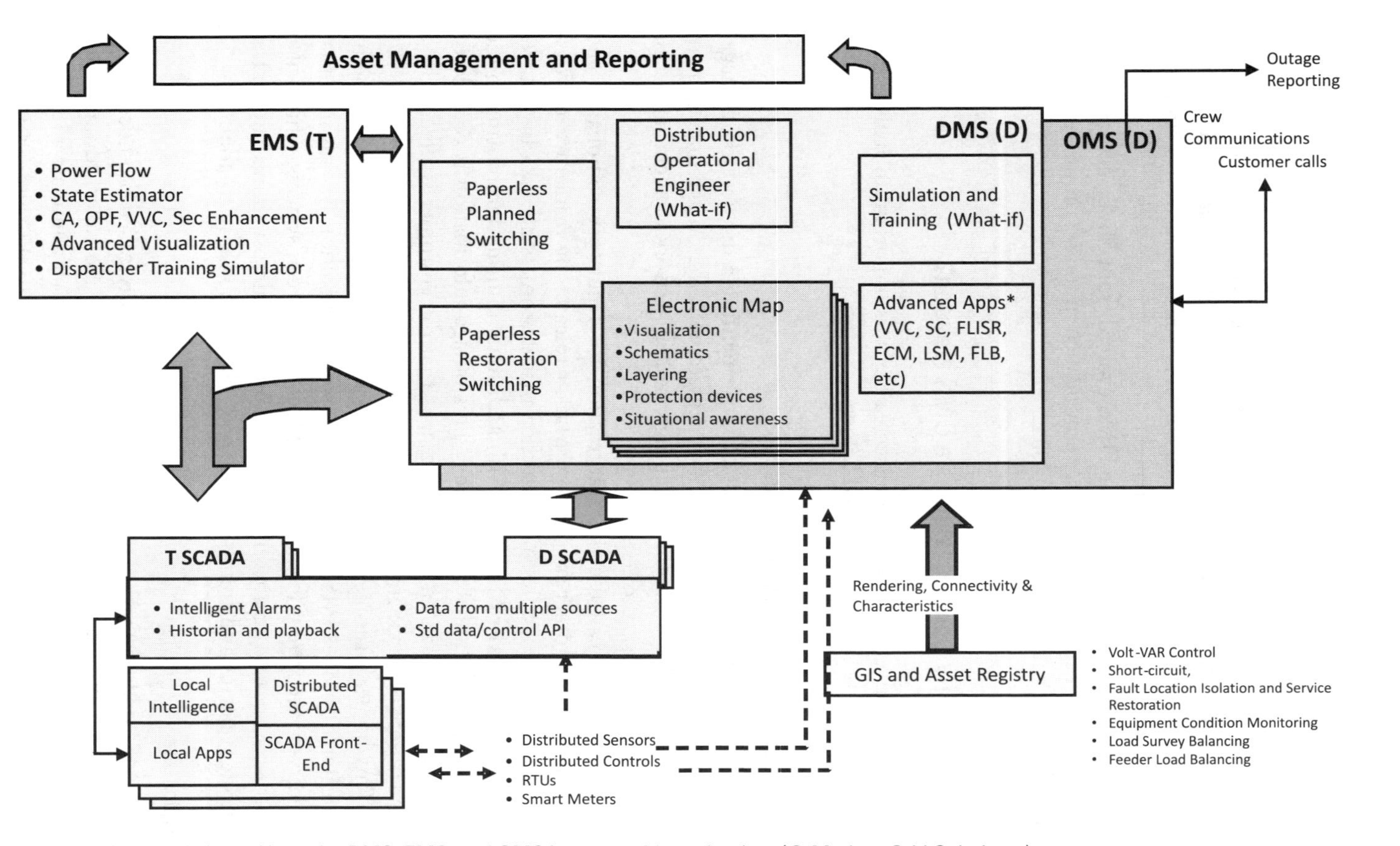

Figure 10.1 High-level view of how the DMS, EMS, and OMS interact with each other. (© Modern Grid Solutions.)

just a SCADA system and evolved to include the network applications and then WAMS, DMS has also evolved over time but in a very different manner.

DMS has evolved from different areas depending upon the strengths of the different vendors who are in this marketplace (see Figure 10.2). This evolution has also resulted in the vendors being able to demonstrate different key factors as their selling points as they have marketed their products to utilities. DMSs have evolved from outage management systems, SCADA dystems, and a combination of SCADA/OMS systems.

While the benefits of where the DMS evolved from can be debated, most of the DMSs are beginning to coalesce around a common set of functionalities, which will be defined later in this chapter.

10.2 The Utility Context: Why is a DMS Needed?

Given the key drivers that were discussed earlier, it is important to analyze how DMS solves the key issues identified below.

Greater Standards for Customer Satisfaction

One of the most important benefits of implementing a distribution management system is DMS's ability to fully integrate with the outage management systems. DMS is able to monitor the progress of all network incidents and their potential impacts on customers.

The operator can use this information for regulatory compliance and improving their customer satisfaction. Market research by regulatory and customer-representative bodies has identified that customers affected by outages are less dissatisfied when provided with additional information relating to the fault and especially its likely duration. This outage information is considered to represent more than an acknowledgment that the customer is affected by a particular network incident.

This higher customer satisfaction factor has resulted in further benefits for integrated DMS, OMS, and interactive voice recognition systems such that the necessary accurate information quickly flows between systems. This is also time-critical because customers contact the utilities almost immediately after the incident begins and continue to do so until after power restoration.

Decision Tools

Most DMS products have out-of-the-box functionality called modules or applications that provide enhanced functionality. They are called advanced applications and give operators tools at their fingertips that automate a lot of the work that would otherwise be manually intensive. This provides the operator

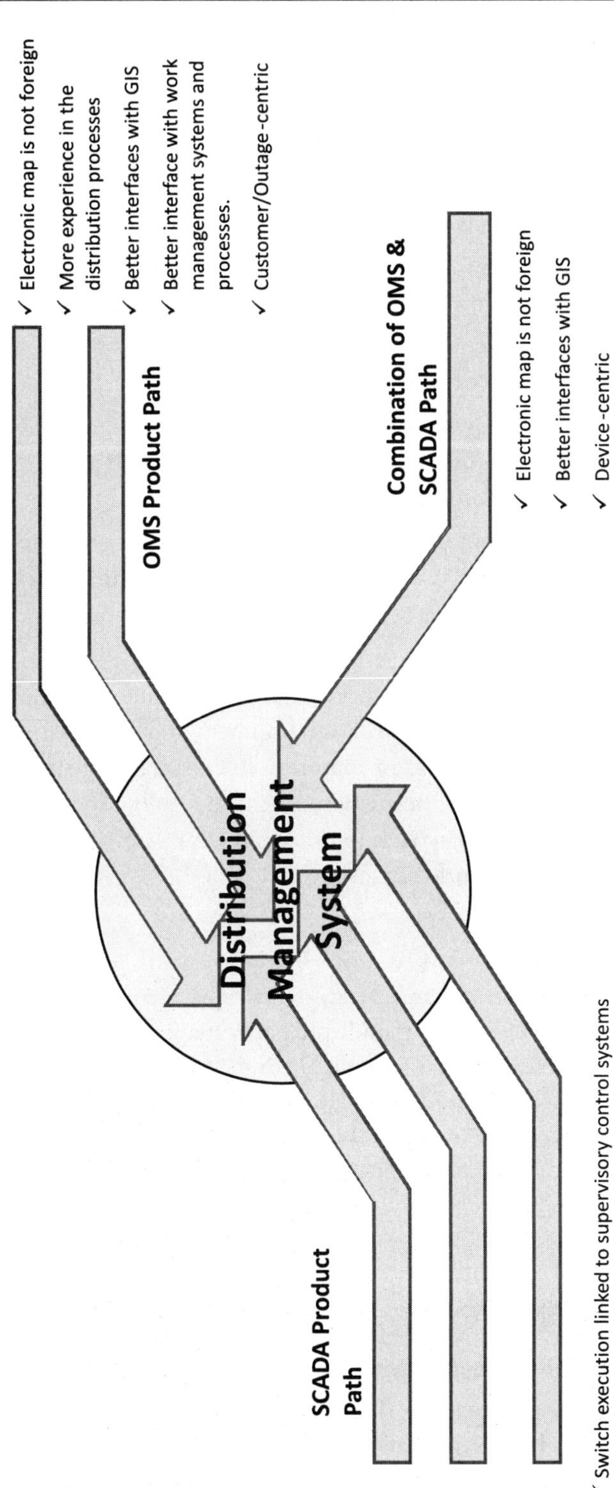

Figure 10.2 Evolution of a distribution management system. (© Modern Grid Solutions.)

with more time to analyze the results instead of spending time crunching data. Typical DMSs tend to include the following advanced applications:

- *Load flow analysis,* which is a real-time load data analysis to accurately represent a variety of network configurations and power system components to relay voltage and VAR analysis and expected voltage impacts.
- *Paperless restoration switching,* which provides a recommended restoration sequence based on real-time and historical data analysis as well as the ability to include potential limiting factor identification and the ability to manually preselect restoration criteria.
- *Paperless planned switching,* which is similar to restoration switching but used for implementing planned outages. Planned outages, as identified earlier, are required to perform maintenance tasks.
- *Study mode,* which is the ability to run switching scenarios in a test mode prior to running the switching steps in the real-world environment. This mechanism also allows for operators to run "what-if" scenarios to maximize switching options.
- *Distribution training simulator,* which is a training module that allows operators to interact with the network environment in a safe manner. DMS vendors have varying levels of maturity in their training simulators, ranging from an advanced test environment to a fully functioning network grid with real-time playback and the ability for a training supervisor to input various real-world scenarios.

Reduced Outage Duration

With a fully integrated DMS, a system operator can expect to manage outages more effectively. Instead of updating multiple paper maps and the outage system, the operator can concentrate on a single DMS map. With a fully functional DMS, the operator can yet standard OMS functions of identifying outage locations, outage extent, and so on, and spend less time creating switching sequences since the DMS can determine and present switching scenarios that best meet the operator's needs.

Moreover, rapid restoration of power following a nondamage fault can often be effected within the threshold time, which requires the incident to be included in SAIFI/SAIDI reporting.

Proactive Management of the Distribution System

Most DMSs also come with a study mode that allows the operators to check for problems in switching steps before they are put into production. They can also test out the various aspects of the daily operations plan prior to agreeing to the final plan of the list of maintenance actions that can be implemented.

Ability to Process Real-Time Data Quickly

DMS and its strong integration with SCADA allow the presentation of important real time information quickly to the user.

For example, in the paper-based environment, if there were an outage, the operator will first need to recognize that an outage has occurred through an OMS or a phone call from a field crew. Then the operator will need to go over to the paper wall map and determine where that outage is on the circuit and the devices that are affected.

In a fully integrated DMS, the outage and deenergized lines would immediately show up on the electronic map in some kind of a visually impactful mode—sometimes as flashing symbols to indicate an outage and colored lines to indicate what section of the grid is potentially without power.

When this happens, the operator can instantly identify the sections with lost power and also the isolating switches that border the outage. Similarly, if there is adequate SCADA coverage, it is possible that the operator could have also identified an overloaded situation ahead of it becoming an outage.

Disaster Recovery

Most DMSs are designed with the ability to have the "as-operated state" of the grid in a centralized location that is replicable to other locations. DMSs tend to be designed with advanced fault-tolerant architectures that can be replicated with other parts of the architecture.

If a control center were to shut down due to a regional/local disaster, control of that jurisdiction can be easily transferred electronically to another functioning center very quickly. The extra load can then be handled by operators in the new location.

Increased Manageability of the Distribution Infrastructure

DMS provides clear and user-friendly interface for visualization of any size section of the distribution network (schematic and geographical), topology analyses with coloring of the network (energization, feeders area, and voltage levels), location of any element of the network (find function), and clear insight in technical data of all elements of the network.

The graphics supported by the systems also allow for the panning and zooming of the displays to view either a larger portion of the distributions system at the same time or a more detailed assessment of a smaller segment of the system. The software can also allow for the conversion of the actual geoaccurate system into a schematic three-line interpretation of a segment of the grid.

DMS is a Tool for Optimizing Employee and System Performance

The advanced set of tools that come with most DMSs supported by strong integration with several systems allow for a combination of improved system and operator performance. The deep integration of technology and processes both on the GIS modeling side as well as on the operational side with OMS, CIS, and other systems allow for serious savings in operator performance due to some possible interesting opportunities:

- Entering data once and having the data propagate automatically to all other systems and processes that require it.
- Ensuring online data acquired in a single space and stored in one place; this should be the same regardless whether the data is coming from the AMI/MDM system or SCADA system or manually entered.

The DMS has several value-added advanced applications that come with the system. A couple of key examples are presented here:

- The Volt-VAR optimization application in any of its forms (VVO, CVVC, IVVC, and others) basically allows for optimal capacitor switching improving the voltage profile between the transformer and the load. Many times this is to reduce energy consumption.
- Fault location isolation and service restoration FLISR allows the operator to automatically identify fault location based on sensors in the field and then use the network connectivity model to identify the most optimum route to service restoration.

10.3 DMS: An Architectural Description

From a macro perspective (see Figure 10.3), the DMS sits the next level down from the EMS. However, from any other perspective, it is a very independent system in terms of its needs to monitor and operate the distribution system. Given that the DMS focused on distribution system, it has some key differences from an EMS (Table 10.1).

10.4 How the DMS Supports the System Operator's Mandate

As time has progressed, the DMS has established itself as the foundation system to support the system operation mandate. It delivers on the mandate through a variety of characteristics:

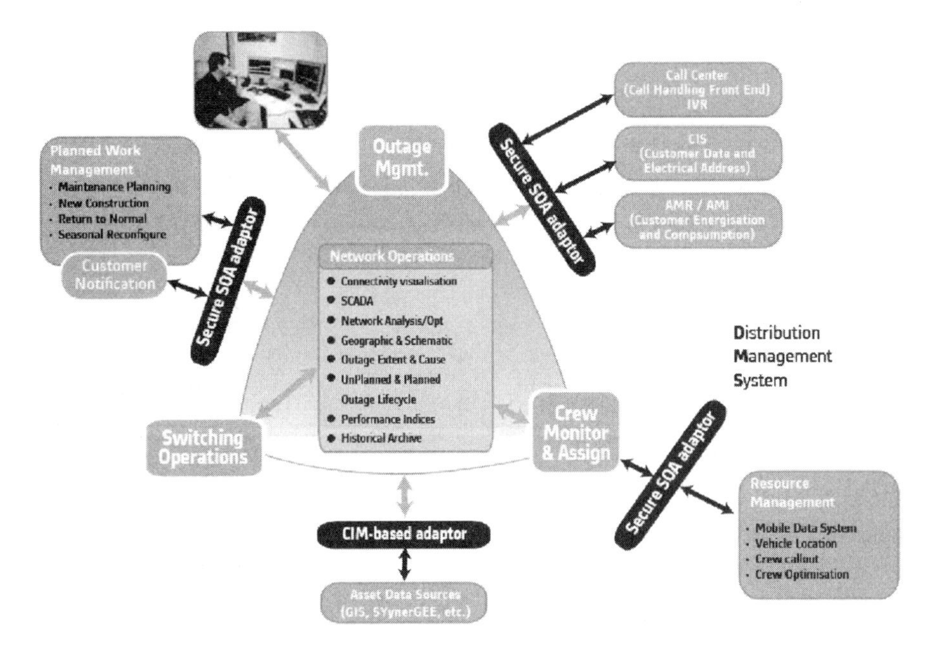

Figure 10.3 Architecture of a DMS. (Picture courtesy of Alstom Grid and used with their permission. Alstom Grid retains all copyrights for this image.)

- *Convergent:* The DMS is the single repository of all operations-related information whether they are the maps/connectivity of the system, asset operational characteristics, limits, and so on. By keeping it all in one place, the DMS has allowed the system operator to go to one place for any specific piece of information related to distribution operations or for the need to take any specific action on the distribution system.
- *Complete:* The DMS has the ability to provide the operator with a holistic picture of the distribution system, taking on information from several different sources, SCADA RTUs, IEDs, substation and distribution automation sensors, field crew location (from their on-board GPS systems), and others. Getting all the data and processing it allows the DMS to provide an as-complete-as-possible picture of the distribution system to the operator.
- *Accurate:* The algorithms used in a DMS provide a high level of precision on their output related to the calculation of the connectivity in the network, systems flows, and voltage magnitudes, planning them on the electronic map, and so on. This accuracy comes from using a sophisticated three-phase unbalanced power-flow as the underlying engine supported by n accurate GIS/power system model and characteristics.

Table 10.1
Difference between the Software Used in an EMS versus DMS*

EMS	DMS
Solves for a single-line system: Even though the transmission system is also three-phase, the phases are generally balanced.	Solves for a three-phase unbalanced system.
The underlying entry point into the EMS is through SCADA measurements, which are then fed into a state estimator that in turn feeds a powerflow to get the final understanding of the flows in the system. This happens because of the fundamental requirement that there are more measurements than states (remember the discussion on state estimator) that it is solving for.	The underlying entry point into the DMS is from the model fed into the DMS from the GIS, which leads to a solved power flow. This of course is the solved three-phase unbalanced power flow. SCADA measurements are then fed into this to update the solved power flow values in the next iteration. This then becomes an ongoing process of solved power flow being updated by SCADA measurements. This mechanism happens because in general this system is expecting fewer SCADA measurements than EMS/transmission—a result of fewer SCADA instances in the field.
The network applications are run in a sequence in the EMS focusing on a full-system analysis all at the same time.	The applications in a DMS are mostly ones that are run as a specific tool necessary for performing specific tasks—switching order creation, clearance management, volt-var control, etc.
Much of the measurements are based out of a substation—again a product of most transmission components being generally all located within a substation.	Much of the measurements are based external to a substation—a product of most distribution components being generally located outside a substation.

*See Figure 10.3.

As a result, the DMS uses SCADA values where available and model variables when actual measurements are not.

- *Real-time:* Given the real-time nature of the DMS and its applications, it has a unique ability to process most inputs extremely fast and also displays that information immediately upon discovery or calculation. This includes not only the calculated values like power flows and circuit breaker statues, but other information like alarms, exceptions, operator logs, and tags are all visible as soon as they are applied.
- *Interactive:* The DMS provides the operator with the ability to interact with the system whether the action is being taken remotely to perform a control action or the operator needs to connect with the crew in the field and work on the control action in a manual manner. This is an important characteristic because many times in switching scenarios, some of the steps could be performed remotely and some steps need the assistance of the field crew. Keeping all the steps and their completeness status (and time stamps) in a single location also allows the DMS to maintain its convergent characteristic.

- *Dynamic:* Most DMSs (see Figure 10.4) provide fairly sophisticated layering and advanced visualization mechanisms as a part of their display systems. They provide the ability to overlay various types of information on the map, like weather data, crew location information, different levels of system connectivity/detail, and so on. The panning and zooming capability allows the DMS to show different levels of detail as you zoom in or out.
- *Advanced:* As they moved away from the SCADA systems running on paper maps, the DMS vendors have provided a platform for advanced applications to run from same real-time or a study model.

  The study model allows the operators or operational planners to run multiple what-if scenarios either on the real-time model or on a future planned model. This is possibly one of the more powerful features of the DMS in that it allows the utility operations personnel to either study a future scenario or learn from a recreated scenario of a past disturbance.

  Incorporating these into a simulation model also allows the operator to be trained on realistic simulations of the distribution grid. These

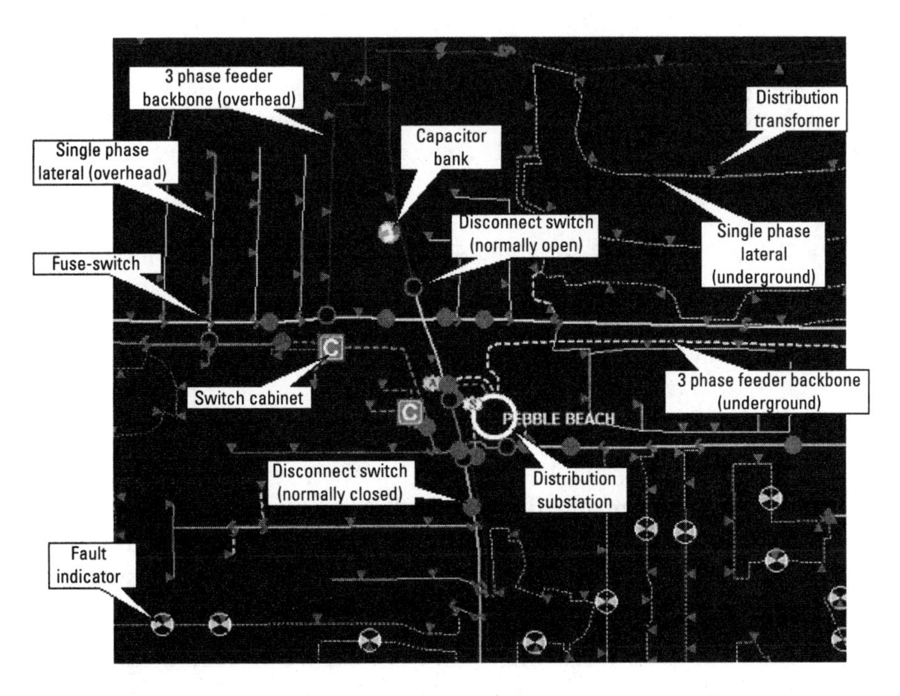

Figure 10.4 Picture of a typical DMS one-line diagram with three-phase and one-phase overlays. (Picture courtesy of Alstom Grid and used with their permission. Alstom Grid retains all copyrights for this image.)

simulation systems allow the distribution operators to be trained on different scenarios in a time-based simulation environment.

- *Ubiquitous:* Until the onset of the DMS, we had discussed the use of paper maps and pushpins as their predecessor. With these capabilities, disaster recovery and backup was almost impossible. The DMS allows the capability to replicate the same information in multiple locations with different levels of permission to perform specific actions. Either upon a disaster scenario or during emergencies when extra hands are needed to help out with multiple nested outages, the DMS allows support operators to provide help all of them functioning from multiple locations. This comes in especially useful to transfer the full control of a jurisdiction from one control center to another either upon the occurrence of a disaster or for work-load balancing.

10.5 How the DMS Supports the Smart Grid

A DMS is in a unique position to support and enable the smart grid by becoming the single repository of all real-time and near-real-time data and power system connectivity model (asset characteristics, connectivity, and renderings). The DMS in conjunction with the EMS and the OMS becomes the basic T&D foundation source of information to support the smart grid.

Three capabilities stand out with respect to how a DMS supports the various mandates of the smart grid. These are in addition to the foundational characteristics identified earlier.

- *Self-healing grid:* From a smart grid perspective, the self-healing grid has always been identified as the Holy Grail. DMS takes us part of the way there. As long as there are sensors in the field that detect faults and send the information back and the DMS has sufficient controls to be able to (either automatically or under manual control) open/close circuit switches, the DMS brings it all together.

  Understanding the location of the fault and combining it with the knowledge of power system connectivity and switch status/remote controllability will allow the DMS to develop a new configuration that will minimize the number of customers lost and also allow the system to be brought back to normalcy in the fastest possible manner.
- *Condition-based maintenance:* The DMS has the best ability to track and maintain records on asset usage and it associated response patterns. This knowledge allows the DMS to analyze and develop different mainte-

nance and/or replacement strategies for distribution assets. Given the study environment in the DMS, it can even try them in a safe environment and run them to failure before determining the best time to maintain or replace them.

- *Predictive monitoring:* Condition-based maintenance is the pre-cursor to predictive monitoring. Understanding usage and response patterns provides the relevant data that will then allows the DMS to analyze usage patterns and predict its possible failure. Doing this in real-time can allow the DMS operator to monitor the situation and take appropriate action based on the urgency, the perceived health of the specific asset, or the system conditions on that day or in the near future.

10.6 Key Component of a DMS

The DMS in many ways is architected in a similar fashion as the EMS. It tends to have several of the same components as the EMS: hardware, software, databases, and user interface (UI).

The following sections will provide insight into these components in more detail (see Figure 10.5).

10.6.1 DMS Hardware

The hardware used in a DMS has many of the same components as the EMS:

- *Computers servers:* Similar to the EMS, the DMS system servers are the main computers on which the various application sets like database services, UI services, mapboard services, and a host of other system services are executed.
- *UI devices:* The UI devices, which are operator consoles, are generally configured similar to an EMS and even other systems within the control center.
- *Electronic mapboard:* Given that DMS systems are newer when compared to EMS systems, the mapboards are directly going towards the electronic versions. But for all other practical purposes they still fulfill the same functionality of presenting an overview of the system state at a high level.
- *RTUs and other front-end devices:* The RTUs and their front-end processors function in a similar way as in the EMS. A newer set of devices that is being rolled out into the field include distribution automation devices

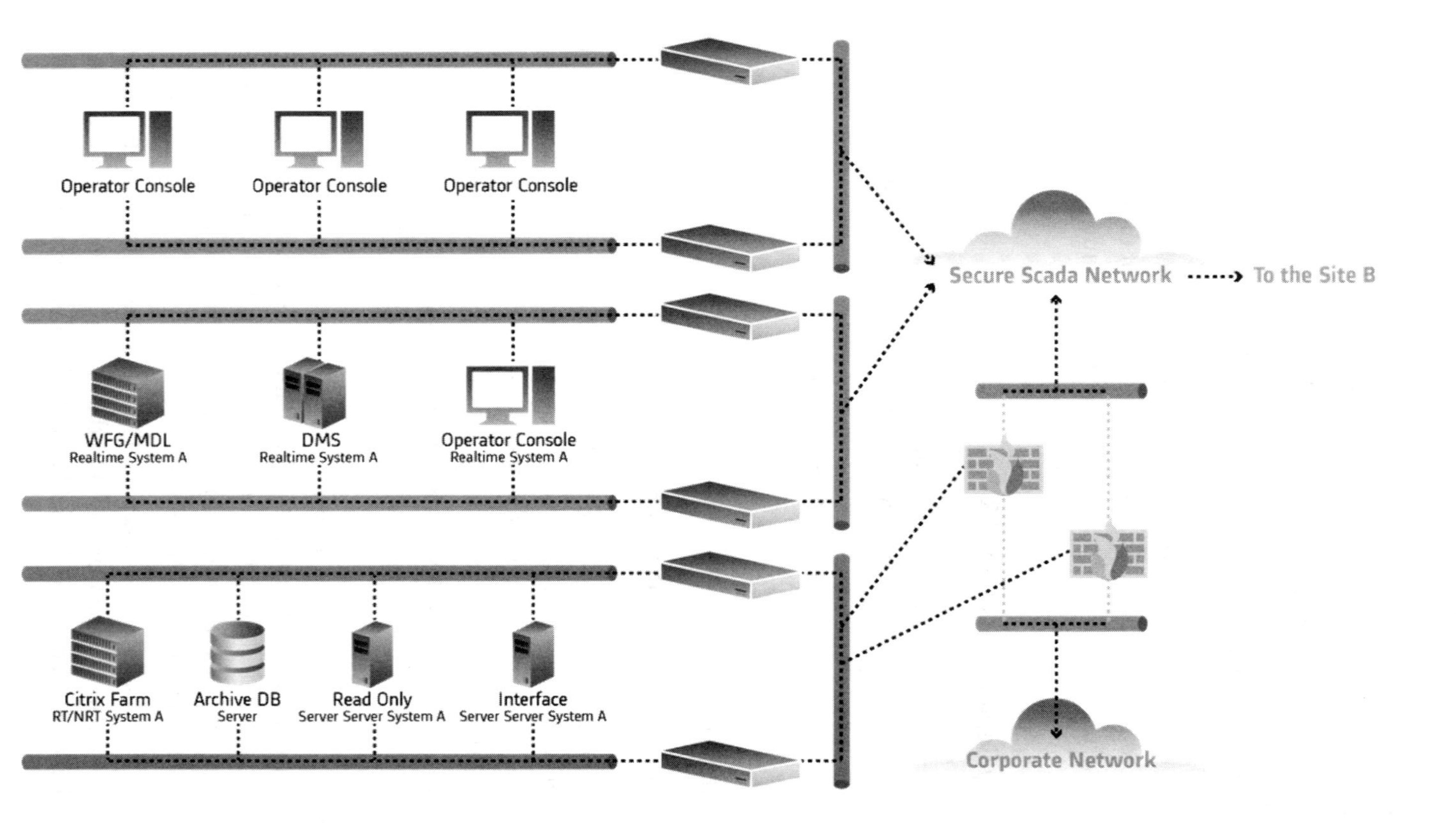

Figure 10.5 Typical DMS technical architecture. (Picture courtesy of Alstom Grid and used with their permission. Alstom Grid retains all copyrights for this image.)

and substation automation devices. While not all need to send information back to the DMS, some do and may utilize mechanisms other than an RTU.

- *Ancillary devices:* The ancillary services devices that support a DMS are very similar to that in an EMS and provide similar services as well—printers and plotters, alarm annunciators, and strip-chart recorders.
- *Communications hardware:* In the communications arena, we have seen the maximum amount of change over that of the EMS. While in the EMS, much of the communications were done over fixed wire to the central site, for the DMS, utilities are even experimenting with cellular-based IP communications mechanisms.

10.6.2 DMS Software

The underlying software used in a DMS is very similar to that in an EMS:

- *Operating system software:* DMSs appear to have moved into very general purpose operating systems like MS-Windows, Linux, Unix, and so on.
- *Application software:* DMSs have a similar set of applicant software that runs the gamut from SCADA (real-time) to advanced network applications to the more business-focused applications.
- *System and utilities software:* While alarms continues to be one of the more important system/utility software in a DMS, some of the other applications that are prevalent in an EMS also tend to be used in a DMS—configuration manager, console, and mapboard control.
- *Network/communications tracking software:* The importance of tracking communications network status while being very similar as in the EMS—it tends to be done in a very different manner in a DMS mainly because of a more diverse set of communications medium being employed here.
- *Process management software:* Similar to an EMS, the DMS also has an overarching piece of software system that manages and controls the execution of all the types of applications, and is called the process manager (or a similar name).

10.6.3 DMS Databases

In the area of databases, DMS applications have moved significantly away from how EMSs were architected. Much of it is architected around relational databases and time-sequenced databases.

10.6.4 DMS UI

The DMS consoles are architected very similar to the newer EMS UI consoles, which are mostly either Windows-based or Linux-based and provide much of the features we are all used to in our PCs and other workstations.

10.7 DMS Application Suites

At the distribution level, utilities tend to want to solve very diffferent problems depending upon where they are located, the legacy of their power system archircture, and what kinds of problems tend to hit them on an ongoing basis. As a result, the DMS application architecture (see Figure 10.6) can be divided into the following three main segments: (1) core applications, (2) distribution and substation automation applications, and (3) integrating applications.

10.7.1 Core Applications

DMSs core applications are design to deliver to the core mandate of distribution operations. This has more to do with supporting the basic day-to-day operations of a distribution utility and less to do with optimization of network

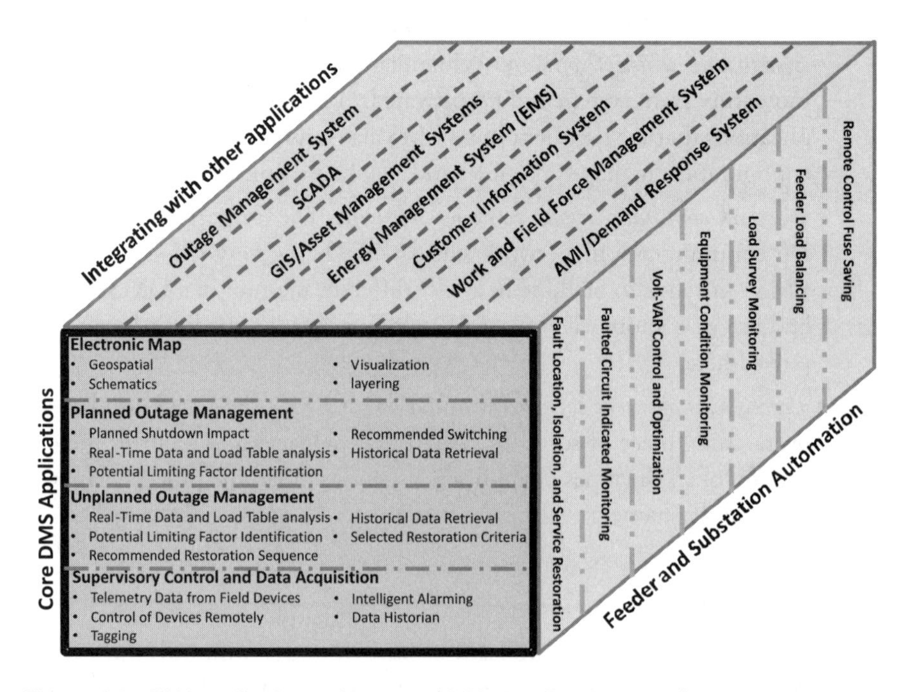

Figure 10.6 DMS application architecture. (© Modern Grid Solutions.)

performance. The latter comes from the addition of distribution automation and substation automation functions.

The core applications of a DMS are:

Distribution SCADA

In general, there is little to no difference between distribution and transmission SCADA systems. The big difference is the need to provide remote monitoring and control of equipment located both in distribution substations and out on the distribution feeders. Just like in transmission, distribution SCADA is a basic building block upon which provides the ability to monitor the distribution system components in real-time or near real time.

The main purpose of any SCADA system is to continuously monitor the loading, status, and performance of equipment located on distribution feeders. Under normal conditions, the system periodically acquires real-time values of current and voltage at various strategic measurement points and the open/closed status of all monitored switches.

Typical sources of DSCADA measurement information and equipment status indications include:

- Intelligent electronic devices (IEDs) associated with distribution system power apparatus. IEDs include protective relays, device controllers (voltage regulators, capacitor bank controllers, etc.), and RTUs.
- Stand-alone sensors, such as faulted circuit indicators (FCIs) and line monitoring devices.
- External systems, such as AMI.

The DSCADA system also enables operators to remotely control devices located on the feeders. Controllable devices include reclosers, capacitor bank switches, voltage regulators, and any other electrically operable device.

DSCADA is also being used to monitor and (in some cases) control the growing number of distributed energy resources (DERs) being connected to the distribution system. DERs include distributed generating units (including renewables) and energy storage units.

Switching Sequence Generator

Besides SCADA, the switching sequence generator is considered to be one of the most important applications within the DMS. This application attempts to deliver to one of the main tasks that consumes the most time of a distribution operator's job whether it is for normal operations or for emergency/storm restoration operations.

The main objective of the switching sequence generator is to support the operator's job of develop a switching sequence to isolate specific components in a power system. During normal operations, it is created to isolate a specific component so that maintenance can be performed on it. During emergency operations, it is created to find the best and quickest way to both isolate a faulted zone and/or to bring power back to the people.

The switching sequence generator works by running thousands of power flow solutions under different switching conditions to identify the one or more sequences that can perform the operation in the best manner possible. Several vendor offerings also provide for sequence generators that can deliver to one or more objective functions: (1) fewest set of operations to perform the isolation or reconnection sequence, and (2) fewest number of customers impacted and others. It also performs this operation on the present state of the network, thereby ensuring that the operation, if triggered, would actually work as planned.

From an implementation perspective, the output of the generator is an ordered list of actions that will need to be performed in sequence. The operator can simply perform the specific operations right from this list itself. It the specific operation is SCADA-enabled, the system connects and sends the control command through SCADA. If it needs to be performed manually by a crew in the field, the operator will communicate to the field crew and record the action as complete as soon as they complete it. This way, there is a full record of the actual completion of the switching action to either isolate or reconnect parts of the distribution grid.

Electronic Map Support

The electronic map is slowly becoming cornerstone of the DMS and distribution operations. Several features of the typical electronic map allows the system operator to become much more efficient

- *Geospatial or schematic visualization:*The operator now has the ability to visualize a distribution network either in a geospatial mode or a schematic mode. The geospatial mode is the actual physical model of the power system in the field with all the components with their state information as they are connected. In a schematic view, the network is depicted in straight lines and at right angles with only power systems components shown. Figure 10.7 presents a DMS map in schematic form.
- *Three-phase or single-line visualization.* The electronic map can visualize the power system in either a single line mode or in a three-phase three-line mode. Each of these modes allows the operator to view the system in different ways.

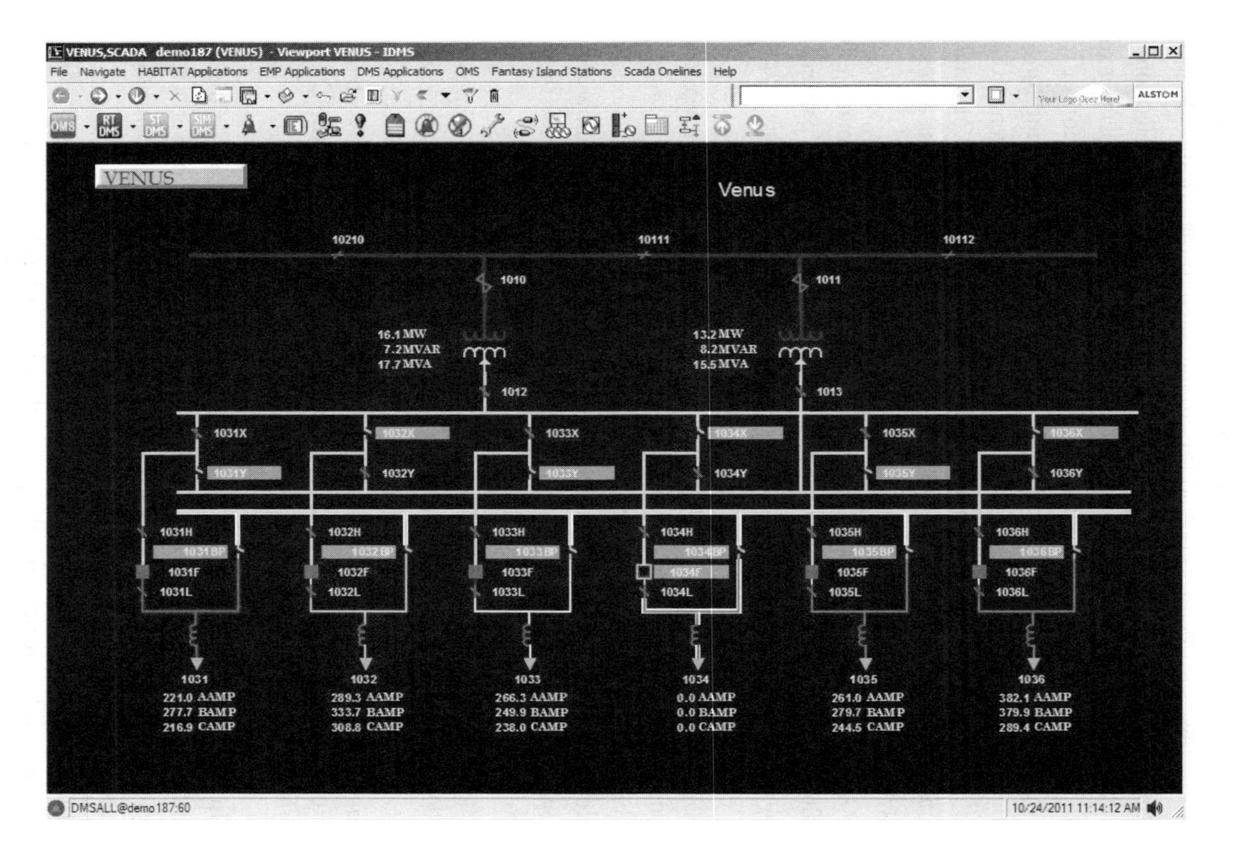

Figure 10.7 DMS output in three-phase schematic form. (Picture courtesy of Alstom Grid and used with their permission. Alstom Grid retains all copyrights for this image.)

- *Providing a topology-based connectivity model.* A topology-based connectivity model allows the operator to look at the distribution in terms of connected/live and outaged segments. The disconnected segment between the two provides the starting point for the faulted locations and also allows the operator to send the trouble-men to that location first.
- *Panning and/or zooming with different levels of details.* The electronic map provides for easy panning of the displays, allowing the operator to follow a specific line or fault all the way to the source substation or to other substations to fix the problem zone. Similarly the zooming function works similar to something like Google Maps in that different levels of detail show up at each zoom level. Depending on the level of detail in the underlying GIS model that is being used, the map can also show details like electric poles, cross arms, and so on.
- *Overlay of other relevant details for the operator.* Most electronic maps also allow for overlaying other nonpower system data like weather patterns, storm patterns, field crew locations, building locations, and street maps/traffic signals.
- *Go up to transmission view or go down to distribution view.* Most vendors support the integration between the DMS and EMS, thereby allowing the operator to go up the chain from distribution to transmission or down from transmission to distribution. This type of feature allows the operator to trace the problem to its original source or even allow them to solve the problem at a higher level.

Distribution 3-Phase Unbalanced Power Flow

A foundational difference in a DMS is the three-phase unbalanced power flow. The sophistication of this algorithm is a differentiator in the quality of the DMS and the fidelity of the system. This is one of the key areas where the DMS differs from the EMS.

- In an EMS, the power flow is solved at the tail end of the state estimator more to solve the state of the system to calculate the flow of power across the various transmission lines. This happens because in a transmission system, there are more measurements than observable states.
- In a DMS, there are fewer observations than actual states. As a result, there is different need for the power flow algorithm here. Here it actually uses the power system model to actually maintain the state, which is then augmented by the measurements that come in from SCADA. Over

time, it is anticipated that as more measurements come in, this model will become more accurate.

The DMS power flow model takes the load model and the connectivity model based on the information that is available to it and solves it.

10.7.2 Distribution and Substation Automation Applications

Distribution automation and substation automation are applications that get added on top of the DMS base applications and very often depend on special sensors in the field or on the substation to support the algorithms. Figure 10.3 shows a subset of them from the field. We will discuss a few of these below.

Volt-VAR Control

Distribution feeders already have capacitor banks that assist in maintaining the voltage profile on the feeder. Very often these capacitor banks are either fixed (no switching) or switched in a schedule resulting in a suboptimal voltage profile. It has been proven that flattening each feeder's voltage profile and lowering the average voltage often results in significant energy savings while simultaneously maintaining unity power factor to eliminate technical losses.

IVVC is considered the most sophisticated of the Volt-VAR control mechanisms and provides all of the benefits of power factor correction, voltage optimization, and condition-based maintenance in a single, optimized package. In addition, IVVC enables conservation voltage reduction (CVR) on a utility's system. CVR is a process by which the utility systematically reduces voltages in its distribution network, resulting in a proportional reduction of load on the network. A 1% reduction in voltage typically results in a 0.5% to 0.7% reduction in load.

In this system, there is generally an optimization engine that focuses on meeting the utility's desired power factor and voltage targets while resolving conflicts. The application evaluates and controls LTC and regulator set points and tap positions, as well as capacitor bank states, in order to maintain target voltages in the distribution grid. It also evaluates and controls capacitor bank states to manage feeder and substation VAR flows. This optimization of resources enables a utility to defer new installation costs by making the most of equipment already in place.

Feeder Load Balancing

Distribution feeders as they are designed [5] in the United States by nature are unbalanced. Depending on the load they are connected to, unbalanced feeders can result in overloaded phases and inadequate use of grid capacity with the extreme case of phase blowouts.

While it is in the interests of utilities to maintain balanced feeder loadings to the extent possible, this is generally only done during initial design of the feeder. After that, there aren't sufficient levels of controls to change the loadings in real time.

However, there are a new set of tools and controls becoming available for feeder load balancing to become a reality. This is done by a combination of using power flow analysis and switching scenario generation and execution to make this a reality.

Fault Location Isolation and Service Restoration

FLISR applications can utilize decentralized, substation, or control center intelligence to locate, isolate, reconfigure, and restore power to healthy sections of a circuit.

FLISR is a distribution automation application that networks groups of switches on a feeder to vastly improve the reliability of utility delivered power by "localizing" outages. Localizing restores power to the majority of an affected circuit, minimizing interruptions to the customers on the faulted portion of the line between the two most local automated switches. The FLISR software can automatically:

- Sense trips (faults) in switches that are monitored and controlled by a SCADA;
- Identify the faulted section;
- Isolate the fault;
- Restore power to customers by automatically switching them to non-faulted sections of the line.

FLISR does not "fix" the problem. Utilities still need to send crews to the site of the fault, where they verify it and make permanent repairs. But many of the customers will experience smaller outage times.

Equipment Condition Monitoring

This is a new area of an extension for the DMS. In the past, asset management at the distribution was considered to be more of an inventory management problem. To support this objective, asset maintenance was, for the most part, done either on a schedule or replacing it when it failed.

With a DMS, for the first time, utilities have equipment usage information. Using this, newer applications for condition monitoring are being designed and developed that can monitor equipment conditions in real time based on usage, peaks/valleys, and cycles. These applications can evaluate equipment condition and inform the operator and even possibly set up maintenance requests as appropriate.

10.7.3 Integrating Applications

For a DMS to function effectively, close integration (see Figure 10.8) with several systems is imperative. While integration is also important to an EMS, it is far more important to the DMS due to the workflow nature of much of the effort that goes through a DMS. In this section, we examine some of the key integration points in a DMS:

Outage Management System

Today, many utilities today utilize their OMS to operate the distribution grid. Connected to a GIS, the outage management electronic map is an effective tool to view the connected assets and associated attributes in the network. However, as reliance on SCADA and distribution/substation automation becomes more prevalent, utilities that manage their distribution operations through a DMS and outages through an OMS end up using two separate systems and need to look at integrating them. For this to happen effectively:

- Operator-commanded SCADA controls need to be passed through to the OMS for it to realize that a specific switch is open on command. Similarly, an uncommanded operation like (possibly) a relay trip also needs to be passed on to the OMS.
- Tags need to be passed on from SCADA to the OMS and possibly back, depending upon where the tag information is recorded. This is in response to a safety issue.
- Manual actions that are performed in response to a crew work order sent out by an OMS need to be tracked in the DMS to ensure that all switch actions manual or remote-controlled need to be tracked in the DMS. Utilities are moving towards a single integrated system which includes both DMS and OMS functionalities.

SCADA

Even though most DMSs come with a SCADA system built-in (sometimes called DSCADA), it is also quite common for DMSs to be integrated with other SCADAs—possibly a SCADA that is more commonly used for monitoring and controlling the transmission system at the same utility.

If the SCADA is an external SCADA, then the interface is generally through an ICCP interface.

Geospatial Information System

DMSs and OMSs get their basic model information from another external system. It can either be a GIS or a homegrown system that delivers the same functionality.

EMS

As mentioned earlier, integration between an EMS and DMS is critical given that they both essentially monitor and support the same grid. For the EMS, the system at the end of a load connection is a distribution grid. Similarly, for a DMS, the system/network that delivers supply into a distribution network is the transmission grid.

This means that the operator, whether it be transmission or distribution operator, needs to be able to go up or down into the other's system just to understand the impact of problems in one propagating into the other. If both the systems are developed by the same vendor, then they tend to have native interfaces between the two. If that is not the case, then the interface is through an ICCP-based mechanism.

AMI/MDM System

While OMSs have AMI/MDM interfaces to provide outage information, the need for AMI/MDM interfaces for a DMS is more of a futuristic need. There is a belief in the industry that given the paucity of sensor information in a distributed system that feeds into a DMS, the information from meters could be of immense value to furthering the improvement of visibility into the distribution network. This subject will be covered in a later section of this chapter.

Customer Information System

The CIS system is the final holder of customer information including consumption and so on. This information is required in the DMS (and if a OMS is integrated into one) to bring the customer side of the equation into the system. Customer information including location and which transformer they are connected to are all information required in the DMS.

Work and Resource Management System

Whether the DMS is integrated to an OMS or not, planned work is still done partly from within the DMS. The mechanism used in the utility to send this information to the field crew is through the work order. While the main work order is created and sent out from within the work and resource management system, a critical set of fundamental information will come from the DMS, switching sequence.

10.8 DMS Models and Its Interface with GIS

The DMS and the supporting electronic map are only accurate as the system model that is contained within it. This model (for the most part) is provided by GIS or another system that could be homegrown and deliver the same functionality. The GIS and its related asset systems provide the foundation to the DMS and as a result, must contain complete and accurate data, strong supporting functions and processes, and a robust, integrated set of systems.

Complete and Accurate Data

The DMS relies heavily on a large amount of data, most of which is typically supplied by a company's GIS.

- *Assets.* Assets include switches (circuit breakers, line reclosers, fuses, etc.), transformers, and other similar components that are operated by a system operator. A key question is "What is the minimum number of assets that are required to model the system?" A utility needs to analyze its goal for the DMS and strategically develop a long-term plan for its GIS. Most times, this would be in addition to the asset management and work management requirements of the GIS.
- *Attributes.* "What characteristics of each asset are required to provide the desired capabilities for the DMS algorithms to solve?" Ratings, location, and engineer/operator notations are just some of the information that is needed to operate off the electronic map (see Figure 10.8).
- *Connectivity.* "How are the assets linked together?" The DMS's electronic maps will turn these connection points into an operable schematic map.
- *Maps/renderings.*"How will I reference the location of assets that are being operated?" Maps typically come from a GIS, which ensures the multiple use of the same graphical interface. Early planning of the maps focused on layering, declutter, symbology, SCADA/control interfaces, standardized naming conventions, layout of devices on maps, and so on, will help ensure a highly usable product for the system operator.

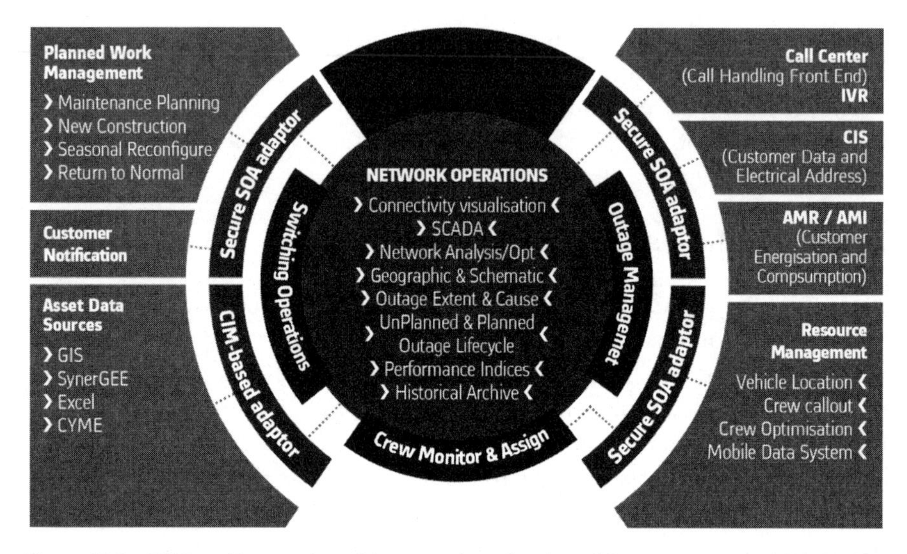

Figure 10.8 DMS and integration with external applications. (Picture courtesy of Alstom Grid and used with their permission. Alstom Grid retains all copyrights for this image.)

Strong Supporting Functions

While utilities generally give primary attention to acquired data quality, they need to consider equally the processes around the management of the data:

- *Data maintenance.* Maintenance processes will allow the various entities that use the data to submit changes (new additions, modifications to equipment ratings, errors identified in the field, etc.) to be sent back in with an expectation of those changes being rolled into the new release of the data model within a predetermined amount of time based on the criticality of the change.
- *Data upload.* The move from paper maps to electronic maps comes with some constraints: it may not be as easy to mark up the maps with the most recent change using a pen or a pencil. As a result, it is very important to have key features like advanced posting of new equipment or configurations for online energizing, more frequent uploads of data (i.e., once a day), incremental uploads versus full data uploads, and so on.

These two processes are important because they determine the accuracy of the data in the DMS and as a result also determine the accuracy of the forthcoming solutions.

Robust Integration

A utility company's GIS and DMS need to be very strongly integrated—efficient transfer of data from GIS to the DMS is required. This is because there are several processes in which these two systems need to be in synch.

- At any point in time in a distribution grid, several changes are constantly taking place. This means that the DMS model also must somewhat stay in synch with these changes as close to the time they are placed and energized in the field. The fewer exceptions to the differences between the systems, the better.
- During storms or other emergencies, it is normal for field personnel to make temporary changes (for example, jumpers) to the system to get the system up and running as soon as possible. During those times, they will come back and implement the more permanent solution at a later time. Until the permanent solution is in place, the temporary fixes must be reflected in the DMS and/or the GIS to ensure that the system being monitored/ controlled is as accurate as possible. During these times, it is considered routine to make the temporary change in the DMS and bring it back into the GIS later.

For these specific reasons, it is important to keep the systems in synch as close to each other as possible. There are emerging standards in the form of common interface model (CIM) for transmission and distribution that is being accepted by most vendors and will enable different parts of the value chain to share information among each other.

10.9 The Future of DMS

Distribution system operations is one of the core functions of a utility. This function entails having primary responsibility and authority for the reliable operation of the electric distribution system. This means they are responsible for the smooth flow of energy to the customer.

The DMS will dramatically change the situation and will be the primary tool of the future to enable the distribution system operator to manage their responsibilities. Its functionality will support monitoring and operating the grid, clearance coordination, switching order creation, and emergency and storm management.

The future of DMS (see Figure 10.9) points toward enhanced capabilities including increased automation, integration with AMI, and smart grid technology. If a utility has not deployed a DMS or is not currently in the process of an

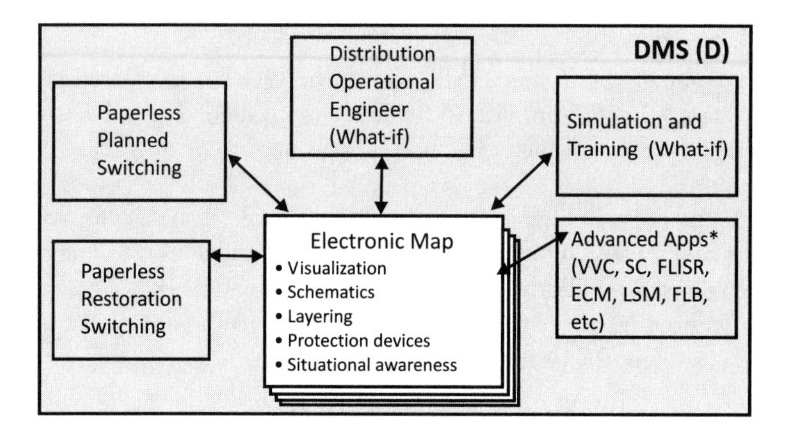

Figure 10.9 DMS and its applications: focus on extensions.

implementation, they run the risk of falling farther behind the rest of their peers in this transformed, highly demanding, and competitive market.

The future of DMS is to incorporate distribution automation (DA) components into its core product. Similar to the way SCADA is widely distributed within the grid today because of the efficiency it provides to the operators, DA components will provide even more in the way of automating the monitoring and control of field devices. This automation gives the operators an enhanced ability to pinpoint outage locations, restore customers and reduce unplanned outage time than what is possible without DA components.

Considering distribution automation (DA), an advanced DMS solution is typically the quickest and most cost-efficient means of implementing a solution. For example, a feeder fault normally requires up to a day to repair and restore power back to the customers. Also, it could take a few hours to restore the electricity to customers that are affected by the circuit switching for this restoration due to manual switching.

While DA cannot reduce the repair time involved, it can provide dramatic reductions to the switching and isolation times with the 3-hour manual switching duration cut to less than 3 to 5 minutes through the addition of a small amount of remote-controlled switchgear. By making use of an advanced DMS, the implementation of DA can be accelerated at lowest cost, providing a solution where the DA intelligence, IT support, and operational expertise can remain at the control center, leaving the remote-controlled devices and simple communications infrastructure in the field.

As utilities integrate more technology to their DMS, they will start laying out the road map towards the smart grid vision. Sensors like SCADA and AMI will monitor events on the grid and provide feedback to operators and engineers in real time. In contrast to today's capabilities, operators will know when and precisely where an outage occurred before customers start calling.

It must be noted that implementing a DMS involves a lot of work that goes into planning, building, and testing the new application. The implementation also focuses a lot of attention on the GIS, a key enabler.

This is only the beginning for electronic maps. Vendors are capitalizing on its potential and are developing the next generation of capabilities. The near future will show us such advances as advanced visualization for operators, providing users with different ways of looking at the same information in a more efficient way.

End Notes

[1] Municipalization happens when a portion of a utility, usually a city with some specific boundaries, decides to secede from a utility and form its own municipal utility. San Francisco, California is an example of an entity that has been trying to (but unsuccessful so far) to secede from PG&E. The city of Boulder, Colorado is an example of a successful secession from Xcel Energy. This just happened in 2011.

[2] 2009 was the first year in a long time where the load growth stayed somewhat flat on average. It is predicted to go up again over the next few years.

[3] Some outages are also caused by either human actions (some in error) or due to inadequate design that results in overloads and hence outages.

[4] Definition of a DMS, © Modern Grid Solutions LLC.

[5] The feeder system that feeds the residential customers is single-phase 12.47-kV network for the most part. This is different from a three-phase residential delivery system in many other countries.

11

Distributed Energy Management System

This chapter is a little futuristic, mainly because this system as specified does not exist but will fulfill a niche that the industry really needs. Many vendors are indeed thinking of a system like this and some actually have systems that fulfill portions of this mandate. Some are more ahead than others but all have gaps. It is not the intent of this chapter to promote one vendor or the other but more to set the vision that (we hope) utilities and vendors can support.

Some of the biggest movements in the smart grid arena are in smart meters, demand response and integration with distributed renewables. Utilities are spending a lot of time and effort in implementing these initiatives and there is a lot of public (and regulatory) attention focused on them. For the first time, utilities are also directly interacting with customers in a way that they have never done before. They are slowly moving from a one-way interaction with the customer into a two-way mode. This is aided by smart meters with two-way communications, time-of-use rates and dynamic pricing, distributed and renewable generation, electric transportation, and storage.

Possibly the biggest impact of the smart grid is in the introduction of new systems into distribution operations. Prior to these systems coming on, distribution operations was either the domain of customer operations or field operations and the main focus was on trouble-call management—which is long considered the precursor of outage management.

Several key changes are impacting distribution operations in a manner quite unlike that of transmission:

- *Pilots.* Thanks to the Smart Grid Investment Grants and Demonstration Grants programs, a tremendous amount of pilot activity has been launched in the utility industry.

- *Distribution automation.* There is a careful and steady move towards installing sensors and controls in the utility's distribution grid with a focus on IVVC, smart regulators, and so on.
- *Smart meters.* It is anticipated that with the end of the Department of Energy (DOE) grants, more than 39 million smart meters will have been implemented in the United States alone. These and their associated head-end systems have resulted in smart meters being treated as a commodity.
- *DMS.* Increased distribution automation is resulting in the design and implementation of operational systems like DMS.
- *OMS.* OMS systems are being implemented either with their functionality assumed within a DMS system or as an independent system.
- *Changes in utility rate structures and advent of demand response:* Utilities are experimenting with time-of-use rates and rolling out new programs to take advantage of them and controlling demand as a result.

These changes have created a level of buzz in the utility industry quite unlike anything we have known in a long time. The diverse needs of utilities are resulting in vendors also jumping in and creating innovative solutions to meet those needs not just in the United States but for a global marketplace.

Even though newer systems like DMS, OMS and others are coming in to the utility operational centers, there is a serious hole in how the utility addresses the customer's needs from an operational perspective. This is because the customer is also making changes in how their energy needs are being met—net-zero buildings and homes, addition of PHEV/PEV to homes and associated charging stations in office buildings, addition of solar/PV cells on homes and offices and so on.

These changes are making an impact on the use and pattern of energy consumption. However, either because these changes have not yet reached critical mass or for some other reason, utilities have not yet provided enough focus on this segment on the management and operation of this segment of their system.

- This is not purely a *customer service issue* but the customer has new and evolving kind of needs that still need to be met.
- This is not purely a *reliability issue* but what the customer can do will impact the stability and reliability of the distribution system.
- This is not purely an *energy supply issue* but the customer can and has the ability to become a net exporter of energy to the grid.

- This is not purely a *regulatory issue* but the outcome of these changes and the need to learn from them is of great interest to the regulator.
- This is not purely a *services issue* but utilities and others who are already providing services to the home (home security, cable, telephone, etc.) are looking at this as an adjacency to identify the possibilities.
- This is not purely a *competitive issue* but unregulated entities could be enticed at defining this market and take more interest in providing new and innovative service to the customers. In doing so, they may move into the utility market and take it away from them.

The author has identified the need for a distributed energy management system that is focused on customer operations, which in turn will provide a be-all and end-all set of operational services to the customer and at the same time allows the utility to manage and operate the grid all the way down to the customer in a safe and effective manner.

11.1 What Is Complicating This Situation?

The present situation, while bringing in a high level of excitement and innovation into a field that really needed some has also created some complicating factors. These complicating factors have prevented the smart grid arena from demonstrating benefits to all the stakeholders in a cohesive manner (see Figure 11.1). They have also prevented utilities from demonstrating substantive benefits to the customer who is the key stakeholder and the regulator who is looking out for them. We have identified a core set of complicating factors:

Data Deluge or Tsunami

There is a tremendous amount of activity at utilities, resulting in a lot of data coming into the utility. Some well-known examples are provided below:

- *Meter data.* There are about 39+ million meters that have either been installed or are in the process of being installed. For the most part, these meters collect data every 15 minutes for each residence. They also collect different types of data for each residence—consumption, voltage (average voltage, min/max voltage, and voltage snapshots), outage data, power quality information, and diagnostic flags (e.g., meter status checks). Several analysts have predicted that, as smart meters get smarter, the data-processing intervals will decrease to around every 30 seconds or so. All this data is time-stamped prior to being sent to the head-end system. This data is generally stored in an MDM system.

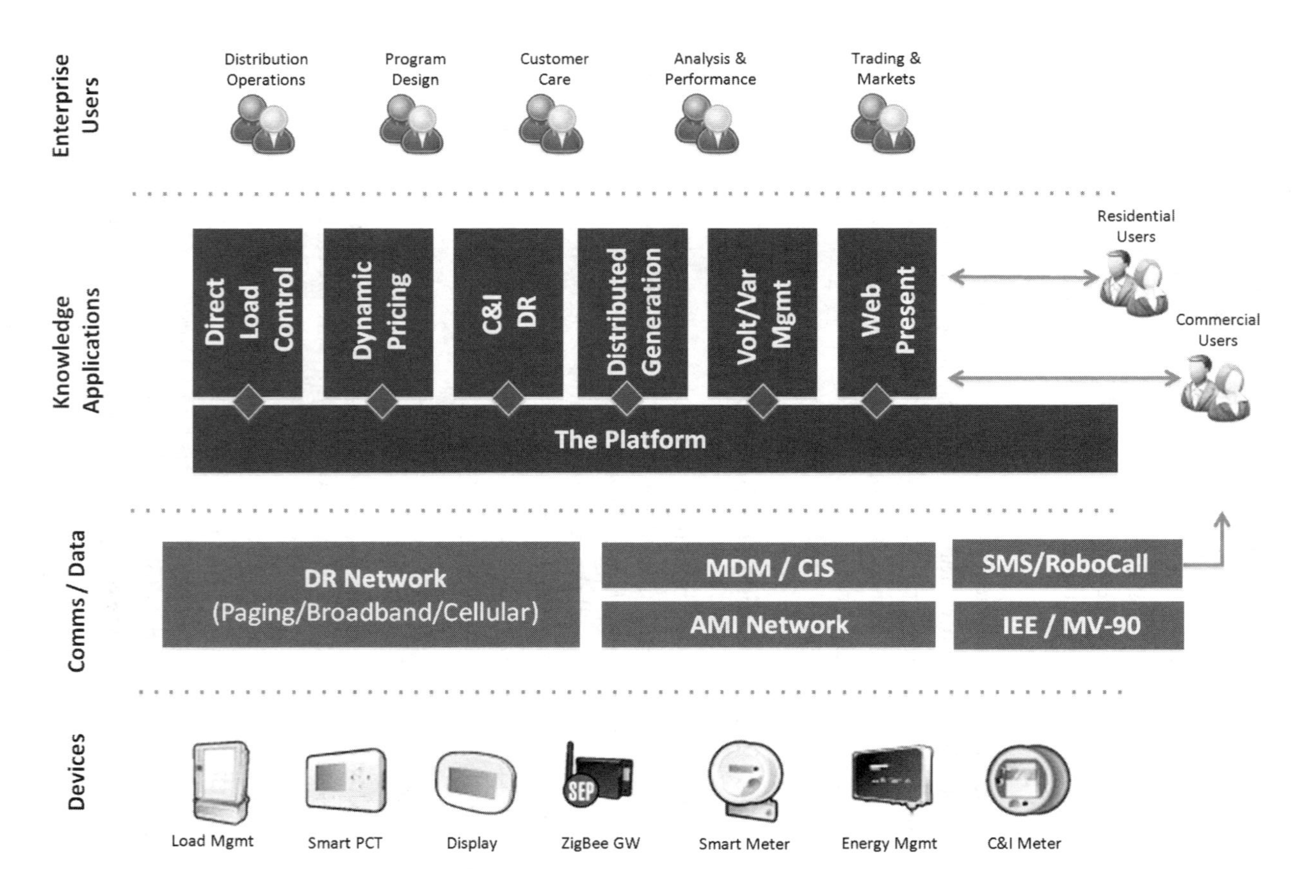

Figure 11.1 Example vendor DEMS architecture. (Courtesy Comverge Inc. All rights are held by Comverge.)

- *Outage data.* An interesting by-product of smart meter implementations is the ability to get a better handle on outages. Outage data is provided as a "last gasp" alarm from most advanced meters. Often utilities can also ping smart meters after an outage to confirm that the power is back on. This detects nested outages, which is when a storm breaks a feeder line in two places, the one closer to the customer is nested within the area affected by the break that is further from the customer. Much of this data is stored in an outage management system and used for managing outages, developing outage metrics (SAIDI, CAIDI, etc.), and for providing information to customers.
- *Substation and distribution automation data.* The smart grid has unleashed a host of distribution and substation automation initiatives which has resulted in several sensors that have been installed on the grid. These sensors, which are in addition to the normal SCADA/RTU sensors that are fed from substation-based potential transformers (PTs) and current transformers (CTs) [1], consist of sensors for volt/VAR control, automatic reclosers, transformer temperature and oil viscosity measurement, fault measurements, and several others.

  Unlike meter measurements, which are taken every 15-minutes or so, these measurements are taken every 2 to 4 to 6 to 10 seconds. Generally all of these come into the DMS system, a substation automation system, or others. These systems hold on to one snapshot of the data for their analysis and store the rest—raw and computed data—in some kind of a hierarchical database.
- *Other sensor data.* With every smart grid pilot or special project, sensors are placed on the grid. The data is brought into some system somewhere in the utility and stored. This could be storage pilots, renewable integration pilots, a specific distribution automation pilots, or others. This data tends to get stored somewhere locally in some system within the utility.
- *Customer program data.* A utility typically runs several customer programs. The programs can either be rate programs for any of classes of customers (residential, commercial, or industrial), demand response programs, energy efficiency programs, or others. For each program, the utility will store information like the details and parameters of the program, the information on the customers who are signed up for it, and the real-time or near-real-time data collected from the participants in the program.

  The information associated with each of these programs is generally stored in the customer information system (CIS). However, it is also not uncommon to store temporary programs and pilots in temporary systems somewhere within a utility and this data may never be migrated into the main CIS system.

The examples provided in this section are just the tip of iceberg of the data tsunami [2] that is expected to hit the electric grid operators. Other examples could include all the information pulled from new smart appliances, electric vehicles, and other metering equipment in the grid. This is nearly a several million-fold increase in daily data processing for most electricity providers—and at this time, they are not prepared for it.

Multiple Noncoordinated Demand Response Programs

Demand response is broadly recognized as one of the early, "low-hanging fruit" from implementation of the smart grid. The potential benefits for managing peak loads—reducing the need for new generation and transmission capacity and mitigating high wholesale power prices—are substantial.

Demand management is garnering attention throughout the electric power industry as a significant mechanism to offset some of this growth by reducing peak demand and allowing the industry's current generation capacity to supply more of customers' required power. As a result, several (if not all of them) are experimenting with some form of demand response However, most utilities are implementing these in a somewhat haphazard manner.

In most utilities each of these programs are being planned and managed independently. This means

- *Programs cannot be managed across each other*, resulting in loss of opportunity depending upon the problem being solved—peak shaving, congestion, etc. For example, if a specific location within the utility is congested, then there is no easy way to identify and trigger all customers participating in all the demand response programs that could potentially assist in relieving the congestion.
- *No coordination of technologies used in these programs.* Most programs at utilities appear to select technology components (at the home, communications and at the utility back office) independent of each other, resulting in a potpourri of technologies with different interfacing mechanisms. While some of this is understandable given the newness of the technology, thought must be given to bringing them all into one single end-to-end technological platform to allow long-term scalability of the program and not have it too dependent on just one technology component or communications technology.

- *Independent tracking of program metrics.* Given the independent tracking of the programs, they are all also tracked independently from a metric perspective. This lack of coordination and the allowing of cross pollination between various programs makes it difficult for utilities to exploit each program to its fullest advantage.
- *Inadequate customer segmentation.* Success of a DR program starts from selecting the right set of customers for each program independent of specifics like location and so on. Very often the customer segmentation is not properly utilized or even performed. As a result, the customer engagement is spotty at best.

Management Reporting

To start, one must understand the amount of activity at electric utilities where a lot of money is being spent on smart grid and related efforts. In what could arguably be called new technology, electric utilities have never spent this kind of money on something new in a very long time.

Given this there is an intense amount of pressure from management, executives and regulators need to (1) show progress, (2) show benefits, (3) show them quickly and in a timely manner, and (4) want the same data in a different manner. However, considering the uncoordinated way that much of this work is being done, creating these reports is slowly becoming a gargantuan task for utility mid-senior-level managers and directors.

Continued Customer Apathy

As mentioned before, this is the first time the utility has tried to monitor information directly from within private residences. The residential customer has for the most part been somewhat immune and left alone from the utility's normal operations activities. The closest the utility came to being inside homes was with the electric meter, which was generally installed when the customer had moved into the home, and after that, no one ever interfered with it. Now, with programs like demand response, the utility is interacting much more with the customer directly.

For an entity that has not interacted with the customers directly until now, there have been a lot of growing pains. Key aspects of those growing pains include:

- *Control/interest in energy at the home.* When it comes to basic energy efficiency upgrades or even energy-related issue, a recent Harris poll found that Americans are still largely apathetic to making changes in their energy use [3]. Whether that is from a lack of knowledge, time, interest, or willingness, utilities have a long way to go.

- *Mistrust.* With a few exceptions, most customers do not either like or trust their utility. The addition of smart meters inside the home allows the utility for the first time to have very detailed information on consumption at the home, leading to privacy issues. This has further exacerbated the situation of trust, leading to several consumers pushing back on smart grid actions by the utility.
- *Opt-in versus opt-out.* Privacy is one of the major issues leading to a movement in the smart meter industry of people wanting to opt out of smart meter installation in their homes.

  A similar issue is also happening in demand response circles. There is a growing movement that says if we need to develop a successful demand response capability at each utility, we need to have more people enrolled. As we identify the people who could be candidates, the thought is to get them all enrolled into the program and have them ask to opt out.
- *Privacy and intrusion.* The change from yesterday's utility where they had a manual read of consumption once a month to a potential place where they have 15-minute (or lesser time) reads in an electronic manner has led to the utility having the ability to understand energy usage profiles within the home.

  In addition, demand response programs at utilities have led to them entering the home for the first time. Devices such as home energy managers, programmable controllable thermostats, and smart appliances have allowed utilities to have an even greater understanding of residential consumption.

  However, this information, if it falls in the wrong hands, can lead to an invasion of privacy.

This has become a serious problem, mainly because much of this effort is focused on the residential customer. Utilities have been working closely with commercial and industrial customers, with special tariffs and energy management equipment at their sites, for a very long time in a mutually beneficial relationship.

Until these problems are resolved quickly, it is possible that the smart grid could either become an expensive experiment in technology looking for a problem to solve, or be taken over by entities other than utilities who may deliver these services to the residential customer, leaving the utility in a purely wires and delivery role. In this chapter, we present the underpinnings of a new system—the distributed energy management system.

11.2 Distributed Energy Management System

Utility operational environments have the benefit of several operational systems like SCADA, EMS, DMS, and OMS. With the exception of OMS, all of the others are primarily power system and engineering systems. This means that their basis at some level is the need/ability to execute a power-flow solution. This requires the involvement of a power system model as the basis of these systems. This requirement brings with it a certain level of sophistication in both understanding and maintaining the power system model and understanding the results that the system delivers. The questions we need to ask are:

- Do we need to go to this level of sophistication for all distribution grid solutions?
- Do we need to solve power system equations to get to integrating and dispatching supply/demand components like demand response, distributed generation and renewables, storage, and/or PHEVs/PEVs?

The answer is NO. There is a place for a simpler system that sits below the DMS. Let us call it a *distributed energy management system (DEMS)* (See Figure 11.2).

The intent of such a system would be to become the singular system capable of managing all distributed energy programs, be device- and system-agnostic, and manage all the data in one place through a common set of user interface into which pilots and programs can be easily be added and removed.

11.2.1 The Core Components of a DEMS

User Interface

The user interface for such a system would be Web-based, focusing on data entry. One could easily visualize several types of data entry that could be supported.

- *Program data entry.* Program/tariff information for demand response, time-of-use rates, regulatory mandates, and limits either instantaneous or cumulative would need to be recorded.
- *Distributed energy devices data entry.* Location and characteristics of all distributed energy devices like homes participating in demand response, solar rooftop PV cells, wind turbines, and microturbines need to be stored in this system to enable them to receive appropriate dispatching commands.

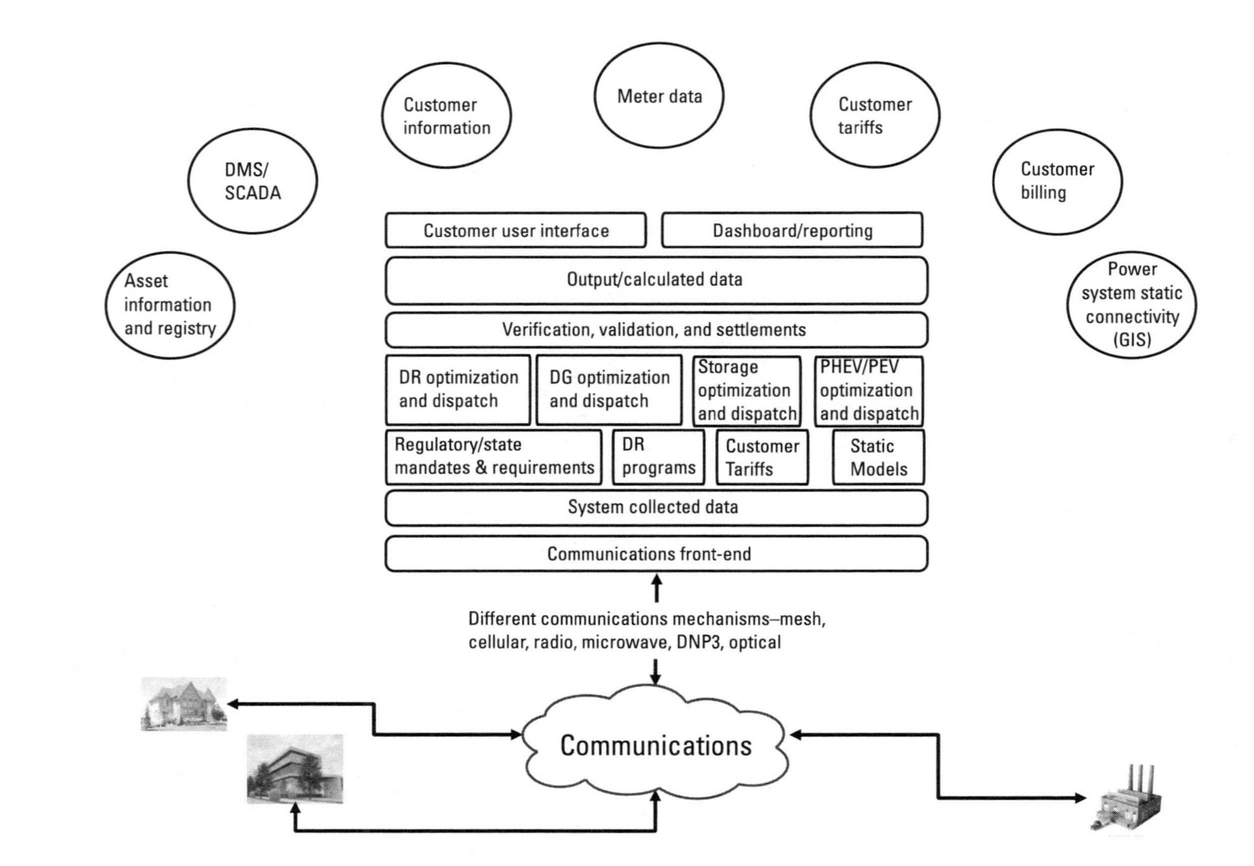

Figure 11.2 An all-inclusive DEMS architecture. (© Modern Grid Solutions.)

On top of manual data entry mechanisms, this system would also support a bulk data upload mechanism from a variety of data formats (CSV, XML, etc.) into the system.

Dashboard and Reporting

Having a top-level dashboard allows for the executive superuser to have a single place to get summary information about every program that is in place and their basic performance as measured by an appropriate metric. This feature supported by a flexible reporting mechanism allows the various users—utility management, program manager, utility executives, regulatory agencies, and so on to quickly get the right data they need and in a form that makes sense to them.

Core Optimization Modules

Depending on the sophistication of the programs being run and/or the need to cross-pollinate between one or more of the programs, multiple types of optimization programs may be run. This could be a core module of the DEMS or a part of the package provided by a vendor delivering one of the demand response program packages. A good example of a program optimization package would be a real-time pricing engine running [4] a double auction mechanism to communicate and control devices at the home based on prices and the homeowner's wishes.

Verification, Validation, and Estimation

Fundamental to managing and operating programs is in validating the outcomes against programmatic information and confirming the benefits to either the utility and/or the customer. This is important mainly from a settlements and reporting perspective.

Programs and Tariffs

Tracking the various programs and tariffs that are being implemented is an important part of the DEMS system. This module not only tracks the programs for demand but also for time-of-use tariffs, renewables, EVs, and so on.

This module will in addition to the program parameters also hold the calculations and algorithms that will be used by the optimization module.

Interfaces

This is possibly one of the most complicated of the core modules in the DEMS system. The core set of interfaces will be with all the systems that have been presented in the architecture diagram in the previous page. The complication comes from the fact that each of these systems is architected differently, is from a different legacy, and ranges from real-time systems to batch-mode systems to database registries.

The interface mechanism needs to be flexible enough to get the right data out of the right system of record to avoid creating a data maintenance nightmare. The possible way to make this work would be through some kind of a service-oriented architecture (SOA)-based architecture that can be flexible enough to support different systems of different types.

Privacy

Whenever customer data is being maintained in a system that is not the customer information system (CIS), privacy concerns take on a certain level of importance of their own. This is because most CIS systems tend to have the right levels of controls in place to ensure the privacy of the customer information is appropriately maintained.

DEMS needs to include a set of privacy mechanisms, protocols, and controls to ensure this information is kept private and cannot be taken out of the system unless personally identifiable information (PII) information is removed.

Cybersecurity

Next to privacy, cybersecurity is an important criterion that needs to be confirmed. DEMS will need to be designed with the appropriate set of cybersecurity firewalls so that data breaches can be caught before they happen.

Now one could argue that there is no need for a separate system such as DEMS. The optimization modules and the dashboard/reporting modules could as easily become applications in a DMS.

11.2.2 So What Makes DEMS a Necessary System?

Fundamental to a DMS is the need for a model capable of solving a power flow solution. This can be complex because the underlying model (coming from a GIS) changes all the time and this model is not an easy one to maintain.

However, if the main purpose of the system is to manage distributed energy sources like distributed generation and renewables, demand response, storage and PHEV/PEV, a solved power flow is not a requirement. It may improve the solution but it clearly not a requirement.

This is an important distinction because irrespective of the user of this system, it brings down the bar to the list of hard requirements that make this system more accessible to utilities of all sizes.

It is also equally important to note that if a utility already has a fully functional DMS, it may be worthwhile to investigate a scenario where the DEMS is either tightly integrated with the DMS or the applications subsumed into one.

11.3 Who Would Use This System?

DEMS would be used by all utilities big and small who have any play in the smart grid space. All have seen the complications identified in this chapter and struggle with its consequences both internally and externally. At a macro level, this is even having an impact on the consumer-perceived benefits of the smart grid and if not taken care of quickly could become serious impediments to the progress in this area.

Large investor-owned utilities (IOUs) who tend to run different pilots across all dimensions of the smart grid could use DEMS to bring the management of all the programs, their data, and their associated reporting under one umbrella, thereby getting the benefits of a single well-designed overarching system.

Similarly, smaller utilities like municipalities, co-ops, and others can use this kind of a mechanism to almost get a DMS lite. It can still allow them to get their SCADA, correlate it with MDM data, support various demand response and time-of-use (TOU) options and bring it all together into one cohesive architecture system. Also, depending on the completeness of their GIS data, they can even use DEMS to project several extra pieces of information on their GIS-based electronic map.

For this kind of a system to be truly effective, it needs to make several inroads into the complications identified earlier in the white paper. Let us see how it can solve the key problems identified in the complications section.

- *Data deluge or tsunami.* With the exception of something like PMU data, DEMS will have the ability to either store the data inside its databases or have the appropriate APIs to get access to the right data within an acceptable time frame.
- *Multiple noncoordinated smart gird pilots/programs in various different technologies* such as demand response, storage, renewables, PHEV/PEVs, and so on. For this to work, DEMS will need to be technology- and vendor-agnostic. Until we can get a full suite of interoperability standards in place, we will need to think of a set of published APIs that will allow DEMS to take part in two-way interconnection with disparate systems. One can easily visualize the need for an interface with a DMS (if one exists) where a DEMS could provide data through a single pipe instead of having every pilot/program having separate interfaces with the DMS.
- *Improved customer engagement.* Having access to all program and customer engagement data in one place will allow the utility to more easily focus on meeting the customer's needs in a coordinated manner, thereby ensuring they are not contacted multiple times for multiple (and pos-

sibly conflicting) programs. Also, having the data in one place will allow the utility to have the best understanding of their response to various stimuli instead of trying to collect all of this information from multiple disparate systems spread across the various back-office computer systems within a utility.

- *Management reporting.* Having all the data in one place also allows the utility to create stakeholder reports that have the ability to go across programs, technologies, locations, and customer classes in a quick and less-laborious manner and be reasonably assured of their accuracy.

Notice that nowhere in this section have we identified the need for a DMS-like power-flow solution.

From a vendor's perspective this is a platform issue (see Figure 11.3). Would they like to be in the hardware/software business or platform business or both? DEMS has the potential to be the first utility-wide system that is truly designed from the ground up to support the smart grid and all its dimensions across the entire value chain.

11.4 Service Models That Need to Be Considered

The movement into smart grid has brought something along with it that is new to the utility industry: Pilots. Until now, most of the systems implementations within a utility have been major procurement efforts—specifications are written, design is performed, RFPs get written and based on it, and the vendor is selected. Then the vendor delivers the solution. This represents a nice clean waterfall-based delivery process. To deliver to this model, utility IT teams have codified strict processes that must be adhered to with respect to design/development of any software that enters their environment. These strict processes have then allowed them to perform routine processes like system upgrades, routine backups, and other maintenance tasks while still managing a system with the right sets of cybersecurity protocols, controls, and access mechanisms.

All of this has changed. There are a lot of pilots, especially in the distributed energy space. Many pilots do not go into full production mode and may not have been even intended to go into full production mode. This means that the systems need to be implemented quickly, get into production quickly, and possibly be dismantled quickly without leaving any kind of a system/digital footprint behind.

For this new kind of a business environment, utilities and their vendors need to come up with new approaches to deliver this capability. Two main approaches are proposed (Table 11.1):

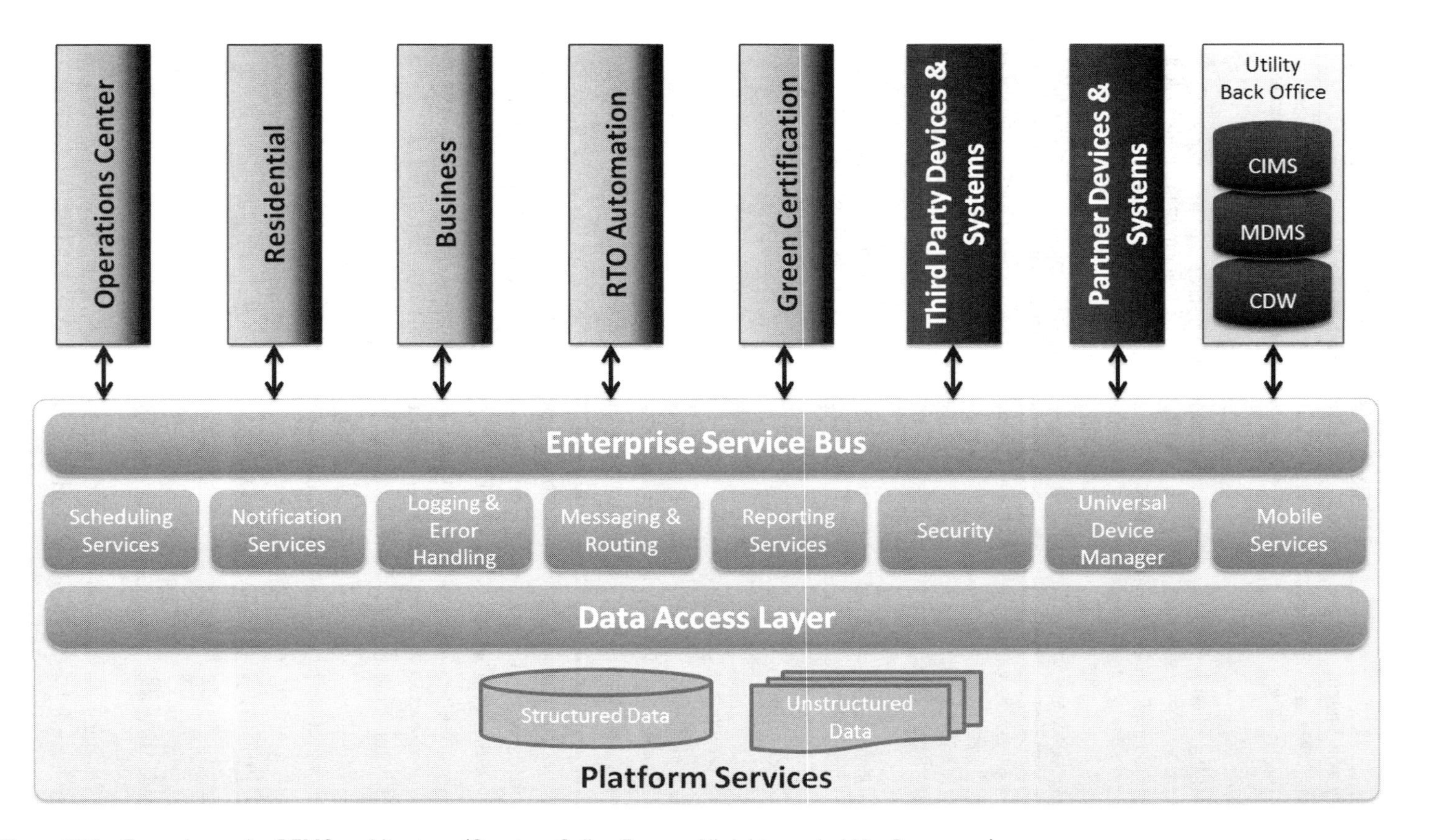

Figure 11.3 Example vendor DEMS architecture. (Courtesy Calico Energy. All rights are held by Comverge.)

Table 11.1
Comparison of Service Delivery Mechanisms

Delivery as a System	Delivery as a Service
System can have customizations as specifically required by the utility.	While system customizations are still possible, it becomes more complicated for the vendor or may impact the overall pricing and hence the associated advantages.
All critical interfaces can stay within the utility's operating environment.	Critical interfaces need to go from inside the utility's operating environment to outside sources.
All system components need to be procured just for the utility—potentially costing more.	Reduced cost: Cost is added incrementally based on actual functionality purchased, saving organizations money.
All storage needs to be purchased specifically for the system—costing more.	Increased storage: Organizations can store more data than on private computer systems.
All software updates, backups, etc. must be done locally.	Highly automated: No longer do IT personnel need to worry about keeping software up to date.
Mobility is something that would need to be added specifically, thereby constraining access to the system.	More mobility: Employees can access information wherever they are, rather than having to remain at their desks.
IT has to maintain the servers and take care of all computing issues.	IT does not have to worry about constant server updates and other computing issues.

1. Delivery as a system;
2. Delivery as a service.

Selecting one versus the other appears to be somewhat subjective in nature. If this question was posed a few years ago, one could have easily said that IOUs and similar-sized utilities would opt for delivery as a system and munis/co-ops and similar sized utilities would opt for delivery as a service. This is mainly because IOUs are fundamentally asset organizations and are compensated for delivery of assets. On the other hand, munis and co-ops are not compensated based on capital expenditure and as a result prefer a mechanism where the cost of the assets can be converted into a monthly rate component.

This has changed with the advent of smart grids and the influx of a tremendous number of pilots across all segments of utilities. As a result of the temporary nature of many of these implementations, even IOUs are seriously considering opting for pilots, because the overhead on making small changes to their IT infrastructure has a big cost to it.

11.5 Challenges

Implementing a system of this kind requires overcoming several challenges. Some are technological in nature and some are business-related.

- *Where does this system reside?* This system is a logical progression away from the EMS (for transmission) to DMS (for distribution) and finally to the customer (DEMS). Given that it is not controlling grid devices, it does not need to reside in a tight access-controlled control center like the EMS and DMS. However, it still performs operations on customer premises by driving the power supply going to residential, commercial, and industrial load by managing the load and supply on the other side of the fence. A better location for this system would probably be in the customer service department in the utility given its closer proximity to customer programs and consumption.
- *Physical and cybersecurity.* At its most fundamental, this system and all that it coves is managing customer data and controlling their energy consumption in some form. While something like the NERC CIP rules do not apply to this system, one must treat this system as having the potential for a medium security threat and apply appropriate cyber and physical security precautions.
- *Privacy and confidentiality.* DEMS is handling a lot of customer data from location (address), location on the grid, usage, appliances at the home that can be controlled remotely, and so on. Standard and applicable privacy and confidentiality rules need to be applied to all the data in this system along with access to it.
- *Interfacing with legacy systems.* As we have identified earlier, several pilots and their systems are being deployed in the utility and we expect DEMS to become the single point of record for all distributed energy related information and the reporting associated with it. This requires DEMS to have two-way interactions with all of these systems. A suggestion for the design would be to create a standardized interfacing mechanism and bring all data in and out through that mechanism; this way the rest of the design and architecture could be managed in a clean manner.
- *How does it get financed and/or paid for?* Given that each pilot is paid for in a separate manner, the common DEMS could be paid for by moving the data entry and reporting costs from each pilot to fund the underlying assets of DEMS. Once the basic underlying architecture is developed, the other modules can be developed over time.
- *Common optimization algorithms.* Given that DEMS would be the umbrella system across multiple pilots, it could also be used to drive common optimization mechanisms as it tries to achieve the best results in energy consumption and still take advantage of new and innovative energy devices like storage, distributed generation, and electric vehicles.

The main challenge here would be in using common algorithms while still trying to achieve the specific mandates of each individual program.

11.6 Does DEMS Have a Future?

The distributed energy management system is a conceptual response to a set of unmet needs. These are needs that are associated with the focus on the customer, which is one of the areas where the greatest change is either happening or poised to take off depending on where in the world one is living. The customers, whether residential or the commercial, will see some tremendous changes happening in the areas of how their consumption is serviced or controlled.

Instead of a different system design to deliver to either a specific type of load or supply, this is an attempt to bridge the gap between DMS and the customer by bringing all of those variations of load and supply into one single integrated system. It is anticipated that this system will grow from being purely an umbrella system over the existing and future pilot systems providing common reporting and model maintenance mechanisms to a full-fledged energy management application suite.

DEMS is in a unique position to support and enable the smart grid by becoming the single repository of all real-time and near-real-time data and power system connectivity model (asset characteristics, connectivity, and renderings) for all utility and nonutility assets leading to the customer and all demand and supply sources within their premise. DEMS in conjunction with EMS, DMS, and OMS becomes the complete suite of applications to support the smart grid.

End Notes

[1] PTs and CTs are capable of sensing voltage and current measurements from the power line to which they are connected. These are generally accurate to SCADA standards and collect data every 2–4 seconds based on need.

[2] Levitt, A., "Riding the Smart Grid Data Tsunami", *Investopedia*, December 23, 2011.

[3] Tweed, K.,"13% of Americans Say They'd Likely Install an Energy Dashboard", *greentechenterprise*, March, 16, 2012, http://www.greentechmedia.com/articles/read/only-13-percent-of-americans-say-theyd-likely-install-an-energy-dashboard/.

[4] Vadari, M., "Active Demand Management—A System Approach to Managing Customer Demand," *Public Utilities Fortnightly*, November 2009.

12

System Operator Training Simulators

The job of the system operator includes some of the most complex activities in the utility. The system operator is one of few sets of 24/7 roles in a utility in which, depending on the circumstances, they are asked to make split-second decisions to solve major problems in the grid, and very often they try to solve problems they may face only once or twice in their entire professional careers. This makes it difficult to develop training curriculum for them because we need to develop an environment that would simulate the real world as best as possible.

For the purposes of this book and the chapter, the term "system operator" has been used in a somewhat generic sense and is intended to include all the people who are involved in supporting the system operations capability. These include the following roles:

- The operator/dispatcher, who sits at the desk managing and operating both the transmission and distribution power systems;
- The shift supervisor, who generally is the senior person on the shift and who leads the group;
- The operations engineers, who provides support through various analysis asks;
- The personnel supporting both the transmission and distribution clearance and switching desks;
- The personnel supporting the scheduling and other dispatching tasks in transmission.

Enter the system operator training simulator—the singular toolset that is designed to support all of them.

The training simulator is a software environment that simulates power system behavior and the (dispatcher) user interface. The simulator may be used for dispatcher training, dispatcher evaluations, engineering studies, power system model evaluations, and offline testing of EMS/DMS/SCADA functions. often, new applications/functions are first tested in the simulator environment before being integrated into the real-time control center arena. The simulator provides a realistic environment for dispatchers to practice operating tasks under normal, emergency, and/or restorative conditions.

Before going into any great details of operator training, it is important for every utility (or its training center) to answer some preliminary questions:

- What is the operator's job in the control center?
- How well does he or she perform the job, and what specific things tell us about his or her performance?
- How can we show these things in an objective manner and help the trainee to overcome any deficiencies?
- How is the job responsibility changing in the short and long term?

From this chapter, one must not assume that a single training simulator exists to train operators all the way from transmission to distribution. Just like there are different systems focusing on transmission (EMS) and distribution (DMS), it is considered fair to assume that there may be two very different training simulators for training those two sets of operators. It is also not uncommon that the two systems are from different vendors.

12.1 Drivers Behind the Need for a System Operator Training Simulator?

The system operations arena has changed dramatically both over time and through deregulation and the smart grid. These changes have also impacted the job of the system operator. As a result of all these changes, the task of training the system operator has become more of a necessity than ever before with today's electric power utilities. What used to be required more for the transmission system operator has now been expanded to include the distribution system operator as well (Figure 12.1).

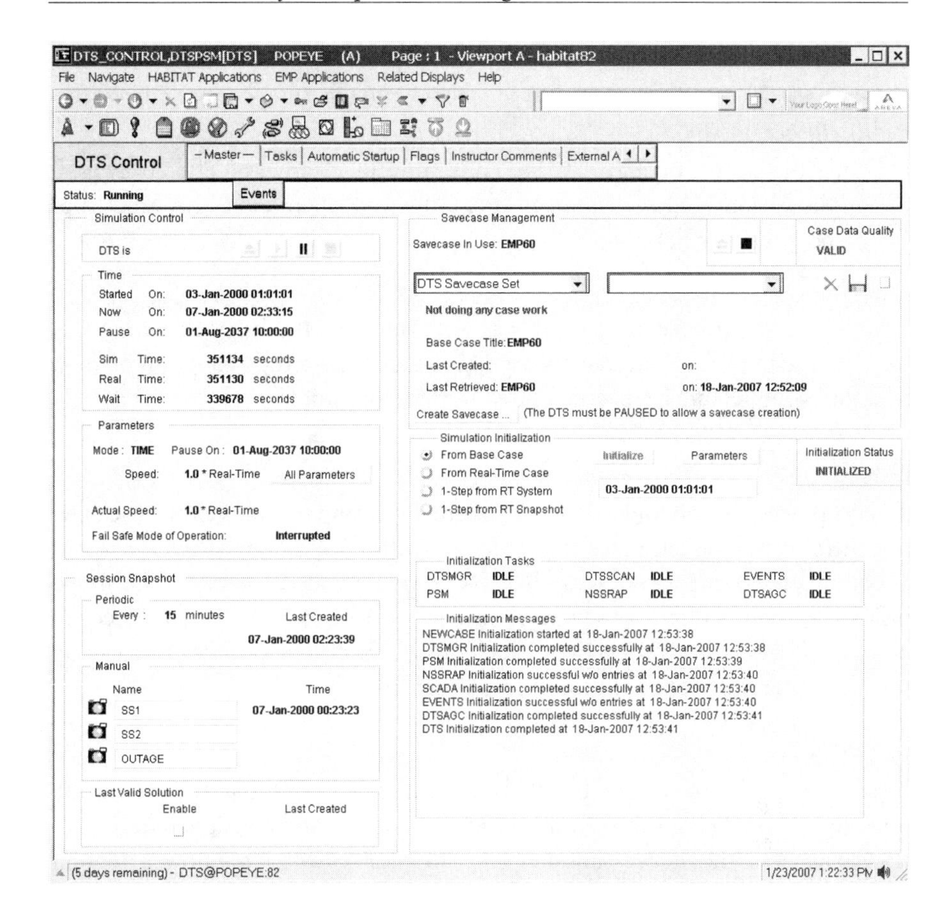

Figure 12.1 Typical main page of a dispatcher training simulator. (Picture courtesy of Alstom Grid and used with their permission. Alstom Grid retains all copyrights for this image.)

There are many factors that have contributed to this situation. These factors have complicated the operator's primary job of operating the system reliably and economically. An operator's performance is generally evaluated by how well he or she runs the power system. However, the following questions probe more fully into the specifics of an operator's performance, and should be considered when evaluating an operator:

- Can the system operator recognize problems and take proper corrective actions in a timely manner?
- Can the system operator recognize opportunities and take advantage of them?
- Does the system operator know the right people to contact for different tasks?

- Does the system operator have the right tools that can be used correctly and efficiently? Are new tools needed to enable the operator to perform more effectively?
- Is the system operator able to communicate clearly and effectively with the various people contacted (field, division operations, and other support personnel?)

There are other considerations as well, but the ones listed above tend to be the more important ones necessary to evaluate from a training point of view. The following factors have contributed to this situation.

New Controls

Today's power systems and their controls are increasing in complexity. The newer control centers include features (such as Volt/VAR dispatch) that were not available a few years ago. As a result, operators now are expected to do much more than their counterparts were doing a few years ago. The operator needs to be trained at understanding these new controls and their possible impacts on the grid under different conditions and scenarios.

Economics

There is a much greater emphasis on economics. For example, the generation operator is now being asked to monitor costs closely and consider various options between either generating the power themselves or buying power in order to meet the demand. Given the advent of power markets and the result of FERC Orders 888/889, economics have played a bigger role in system operations and changed the scenario from cost-based to price/bid-based selections for both energy and all types of ancillary services.

Security of the Changing System

Many utilities are finding that their generation and transmission facilities are not keeping pace with their increases in load. As a result, more and more power is being wheeled across longer distances, very often through congested corridors. The power system is being strained in ways that were unforeseen a few years ago. Utilities are operating closer to their steady state and dynamic security limits. This is even more of a concern as DMS systems are being rolled out, because this is completely changing the entire system operator paradigm for the distribution operator—moving them away from paper-based manually intensive operations to more levels of automation.

Safety Concerns

On an ongoing basis, operators work closely with various field service personnel like linemen, trouble-men, and field crews on bucket trucks. This combine team works on isolating field equipment, performing actions on the equipment, and restoring them back to service. Operators need to be trained on the various switching and operating actions which they perform to ensure that they are fully trained on all aspects of safety.

Regulatory Changes

Between transmission and distribution, there are a plethora of regulatory agencies that drive the business rules and reporting requirements needed to be followed by the system operator. Regulatory agencies such as NERC and local PUCs have set up power system operating guidelines that should be followed for proper system operations. In addition, with the advent of the smart grid, there are numerous new and ongoing regulations that keep coming on a fairly regular basis (DR [1] pilots, TOU rates, RPS mandates, etc.). It is imperative that the operator understand what is required for compliance with these rules.

Smart Grid

The advent of the smart grid has brought several new changes to the grid and most of these changes are either directly or indirectly impacting the job of the system operator. New and more exotic types of sensors and control are going on the grid, new and special types of tariffs are being created, new and complex reports are being required, and much more. For example, when new sensors go on the grid, the system operator needs to understand (1) how the sensors work, (2) what the data actually means and how they should react to the extremes of the data, and (3) how to support the field crew when they are either installing, removing, or maintaining the sensors.

Renewable Power Generation

After smart grids, distributed sources of generation from wind and solar are some of the changes that are impacting the stability of the system the most. Transmission-level wind farms are almost considered one of the most significant nightmare scenarios for the system operator.

For example, large wind farms with generating capacities of the order of thousands of megawatts can go from zero (or low) output to full capacity output within minutes. The same can happen in reverse as well. When this happens, the system operator has the responsibility to scramble quickly and use the levers available to them and rebalance the generation. They need to be trained to do this—otherwise, it can lead the system closer to instability.

In distribution, a similar issue can happen with solar PV installations on homes or through PHEVs/PEVs at the residence. During the middle of the day, when solar PVs are generating a lot of power, it can result in a reverse flow of power situation in a system fundamentally designed for one-way flow of power. Besides the obvious safety issues that can come from the two-way flow of power that needs to be taken care of, it is also causing newer and more complex problems in the system that need to be solved by the operator, who needs to be trained for this as well.

Self-Healing Grid

Most system operators will tell you that they are OK with having relays in the grid because they help to protect and this is done by opening breakers/switches in the event a problem is detected. Now, for the first time, the concepts of the self-healing grid are taking hold, meaning that switches may be opened/closed automatically. The author is not sure as to how far this concept has been taken, but one can already see evidence of this in the areas of feeder load rebalancing, FLISR applications, and so on. When switches close automatically, safety issues become more paramount and appropriate new process steps may need to be taken. This brings in a new level of complexity to the system operator's life and leads to an increased requirement for training.

12.2 What Are the Key Characteristics of a Good System Operator Training Simulator

Besides the obvious characteristics of any software system that it be easy to use, have a user interface that is intuitive, and so on, a system operator training simulator needs to have a specific set of characteristics that make it a good simulator for use in training system operators:

- *An exact replica of the system* used in the real control center. When a system operator is being trained, whether this is for transmission or distribution, it is important that they still use the same set of tools that are normally available to them in the control center at their desk. So a good training simulator needs to maintain an exact replica of their system along with user interface and actions. For them it must appear as if this is the real power system that they are controlling.
- *An instructional position,* where the simulation can be set up and monitored/ controlled. Any simulation environment, especially one in which a very complex system is being managed, needs a control panel from where the simulation is controlled. This includes starting/stopping the simulation, selecting the right save-cases that drive the scenarios for the training, and so on. Generally the instructor tends to be one of the more senior dispatchers or possibly a retired dispatcher who understands the

simulation/training objectives. The simulation metrics are also generally gathered here to support the eventual evaluation of the training outcomes.

- *An event subsystem.* Fundamental to a training simulator is the ability to execute events in the power system. An operator in their normal job responds to incidents in the field. Most often, they do not know the core problem. They start from a series of alarms pointing to some abnormal events in the system. From this, they need to diagnose the core problem and bring the system back to normal. This same aspect needs to be simulated through an events subsystem, where the instructor can perform actions in the power system like trip an overcurrent relay, open a circuit breaker, or trip a generator, thereby taking the system away from normal and observing the operator respond to this stimulus. Very often in most simulators, events can either be deterministic, conditional, or probabilistic.
- *Simulation fidelity* must of a high level. The simulation fidelity comes from having an excellent model of the various components in the power system down to the right level of detail. Power system components that drive the fidelity, meaning the components whose models need to be paid more attention include generators and associated turbines and boiler subsystems, load models, and transmission and distributions lines and transformers. The solving of these models drives the automatic generation of the right response to various stimuli (events).
- *Ease in manipulating/configuring the system.* Training scenarios are generally given multiple times either to the same participant or to multiple participants (e.g., all the operators in a utility). To support this, the simulator needs to be able to support different operating scenarios multiple times. Some kind of a save-case mechanism will allow the instructor to bring a combination of power system model and the events associated with it together to drive a specific training session.

  Adding on to the ease in manipulating and configuring the system is the ability to reinitialize the simulation and start all over with no memory of the previous run of the same simulation. This is an important aspect of a good training simulator because it allows the instructor to test the scenario multiple times before confirming that it is ready for use for training purposes. Sometimes, this also comes with the ability to run the simulation faster (or slower) than real time.
- *The ability to simulate and learn from external conditions.* A very common occurrence in a control center is in facing a new problem that has never

been faced before. This could be a major storm, a sequence of events which happened in the most improbable manner possible, or a generator tripping during the most inopportune moment. When anything like this happens, most EMS/DMS systems have the ability to store all relevant information associated with the event. A good training simulator would have the capability to initialize from this event and use it to train the operators to learn from both the operator's response and also when went right or wrong (see Figure 12.2).

- *A real control center environment.* Beyond the capabilities of the training simulator there is one key point that very often tends to get missed in developing a good training program. There is a need to re-create as close as possible the actual control center environment all the way to console/desk/monitors/phones and so on to create a realistic simulation of the actual response to actions. Re-creating the control center environment

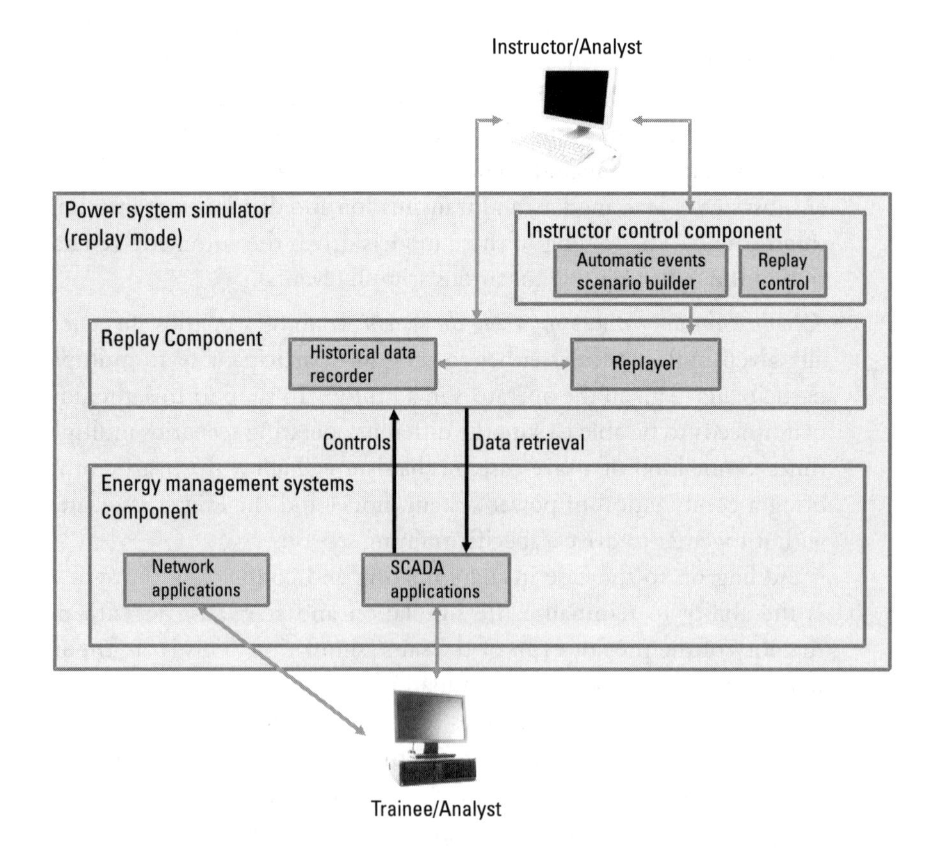

Figure 12.2 Simulator in replay mode. (Picture courtesy of Alstom Grid and used with their permission. Alstom Grid retains all copyrights for this image.)

provides the seriousness of the training and ensures that the operators who are getting trained are getting the full picture.

12.3 Architecture of a System Operator Simulator

The training simulator is generally comprised of three main subsystems:

1. *The control center model.* This is generally an exact replica of the control center applications set that the operator would normally be interacting with on a normal day-to-day basis and would include all the normal applications like SCADA, alarms, mapboard (sometimes), and the set of advanced applications. This subsystem is purely configured to be the view for the operator who is getting trained.
2. *Instructional subsystem.* This subsystem is designed and developed for the instructor of the training program. This is where the instructor configures the simulator to set up the training session: the power system model is configured to deliver to a specific training, the events are set up, and the overall environment is made ready. This is the also the environment used by the instructor to monitor and control the simulation during the actual training session to create as realistic-as-possible experience for the operator (see Figure 12.3).
3. *Power system dynamic simulation model.* The power system simulation subsystem is the most critical part of a training simulator. In the real world, actual transformers, circuit breakers, and loads exist to provide SCADA data for the EMS/DMS and so on. However, in a simulation system, we need realistic models of all components that are generally in the field. These models need to be realistic enough to ensure that their responses to operator or system stimuli are such that they mimic the real component. They also must create the right kinds of data to be fed into the SCADA subsystem because that s the only way data comes into the EMS/DMS (see Figure 12.4).

12.4 Key Challenges in Setting Up a System Operator Training Environment

There are three key challenges in setting up a system operator training environment: (1) Hardware/software environment along with the necessary interfaces, (2) the database models and event scenarios, and (3) the actual training environment. Let us analyze them one by one.

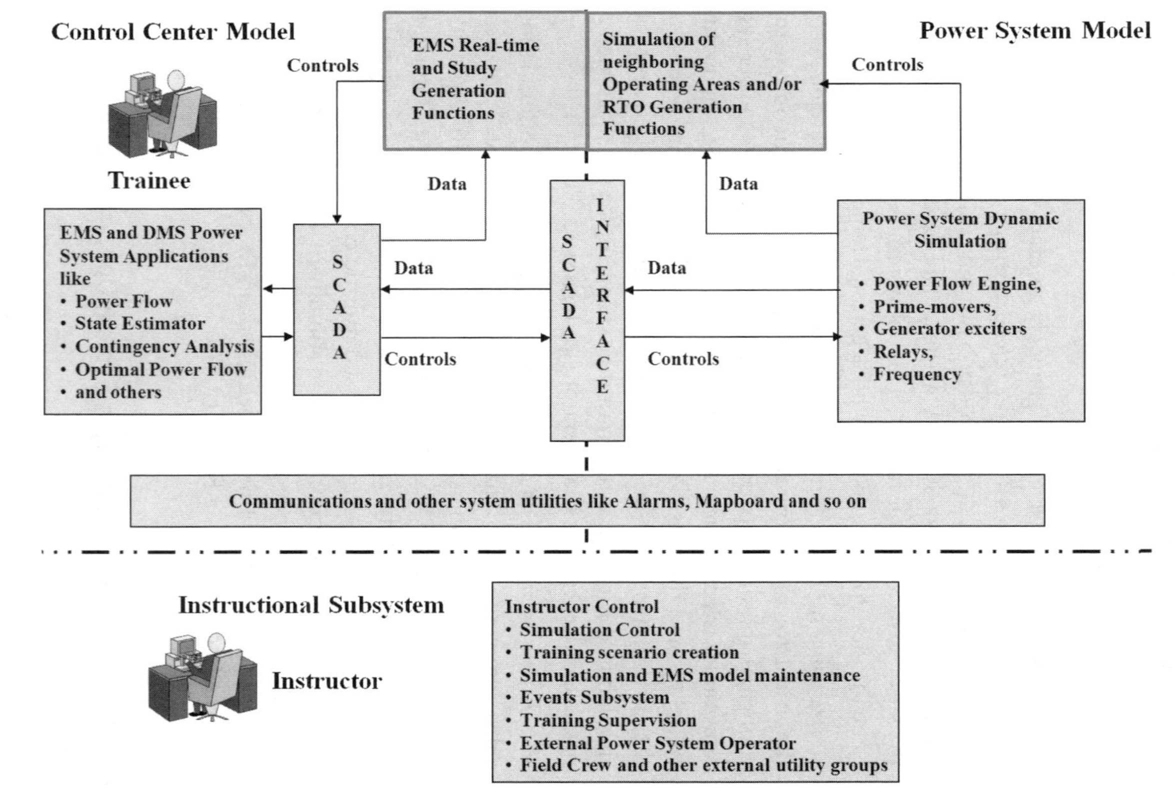

Figure 12.3 Distribution training simulator architecture. (Picture courtesy of Alstom Grid and used with their permission. Alstom Grid retains all copyrights for this image.)

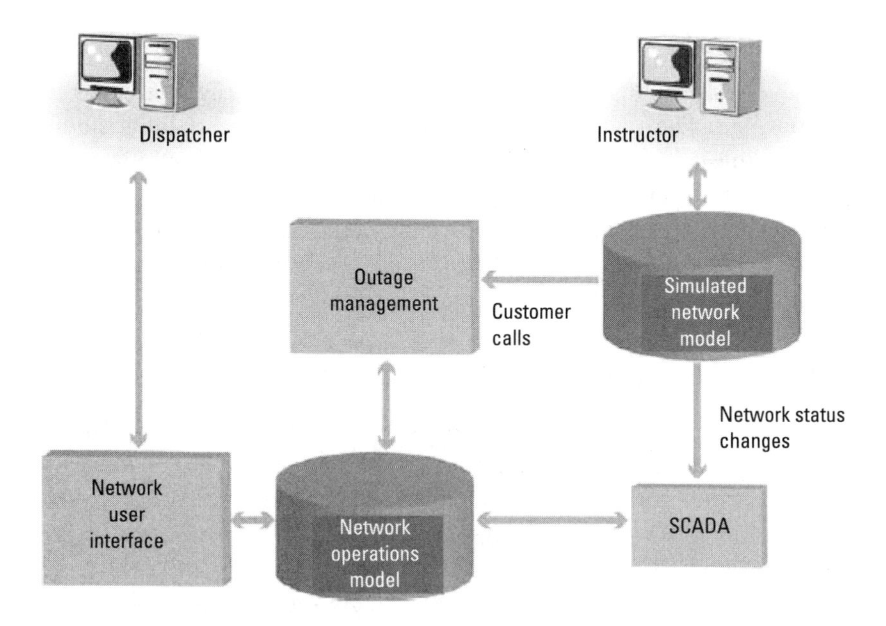

Figure 12.4 Architecture of a distribution operator training simulator. (Picture courtesy of Alstom Grid and used with their permission. Alstom Grid retains all copyrights for this image.)

12.4.1 Hardware/Software Environment

Depending on the compatibility of the software systems between the EMS/DMS vendor and the simulation vendor (either same vendor or different vendors), this can be either a simple job or a reasonably complex job. Most utilities tend to procure the training simulator from the same vendor as the original EMS or DMS system.

As identified in the architecture picture, there are three main systems in a training simulator: control center subsystem, simulation subsystem, and instructional subsystem. Depending on how it is configured, these three subsystems can either be in different computers or all in one.

For the most part, the training simulator is a fairly self-contained system. An exception to this is the interface between the real-time system (EMS or DMS) and the training simulator to allow after-the-fact study/analysis of key events and associated training to operators to learn how to react to them. This interface could be either a real-time interface to move the saved data from the real-time system to the training simulator and simulate the actual event right away. This interface could also be a simple mechanism to transfer save-cases across as a file transfer to the training simulator.

The challenge here is to maintain the systems and the interface with real-time systems in peak operating conditions so that the systems function all the time in a seamless manner.

12.4.2 Training Environment

Good operator training centers tend to re-create the real-time system as closely as possible. The training environment could include a training simulator room(s)—one for the trainee operators and one for the instructor, operator console desks, phone systems, mapboards, and so on.

The main challenge here is in the ability to justify the expense to purchase and integrate these items into a fully functional training environment. The next challenge is to maintain these systems on an ongoing basis.

12.4.3 Database Models

The biggest and most critical challenge in setting up a training simulator is to find the data and the settings for the various models that are used in a simulator. As described earlier in this chapter, these simulators can have rather sophisticated models of key power system components like generators (turbines, boilers, and exciters, for hydro, coal-fired, gas-fired, nuclear, etc.), relays, load/consumption of power, and so on. The data and the settings for these models are not easily available and it obvious that the fidelity of the simulation is quite heavily based on the quality of the data going into the models. Bad modeling provides inaccurate simulation results and thereby has minimal value for training. When simulators are first set up, it is extremely common to have a team of people working painstakingly to find the model and settings data for these devices by combing through manufacturer documentation, IEEE reports, and other places for the information and validate them internally with experts before entering them in.

Another key challenge in setting up an ongoing management of the training program is in the maintenance of the database models. A training simulator generally has multiple databases that need to have a lot of care and feeding. The power system models can have a time-period aspect to them as they could change over time and new components could get added or removed. Keeping them all synchronized require time and effort on the part of one or more personnel.

12.5 Key Challenges in Setting Up a System Operator Training Program

Most utilities would agree that the job of their system operators is very critical to the stability of their utility's smooth operations. In transmission, bad decisions can impact various things from blown equipment to the stability of the overall grid. Similarly, in distribution, bad decisions can impact everything from customers being out of power for too long to causation of safety issues for field crew. However, most utilities tend to step back and think carefully before spending money to obtain the right tools and training for those same operators. This is generally less of an issue for training transmission operators than for distribution operators. It all comes back to a discussion on *people, process, and technology*. How much one should spend on these items varies widely from utility to utility.

People

Utilities need to make a decision on how much training is required and what kind of training their system operators need. This will drive the need to identify the level of formality of training required for those operators and other support personnel. A typical training program for transmission operators alone requires a staff of about 4 to 5 people. If you add distribution operator training to this list, one can easily see this go as high as 10+ people or higher depending on the number of operators to train. Utilities need to balance this cost against their priorities regarding spending for training.

Process

As new system operators come on the job or when something has changed in the field (a new type of control such as IVVC) or in the EMS/DMS/SCADA system, then utilities need to develop a process around how to get their operators trained to implement them. This obviously is more critical for a new operator but major changes can also throw experienced operators off their norm because the power system may no longer behave in a manner they are used to.

All of this needs a very formal and structured process for utilities to ensure that their operators are trained prior to handling the controls of the system. This can go the entire continuum of ensuring their operators are fully trained prior to managing the system on one end to training the operator on the job by pairing them up with more experienced operators and moving them along the learning curve.

A typical set of activities used to train a transmission system operator consist of a structured set of areas they are exposed to and trained, all the way from orientation to full analysis of the response of the system under normal and emergency conditions.

An illustrative set of examples are identified in Table 12.1.

Technology

Procuring a viable simulation system is only the start to establishing a full-fledged training program. The technology can go from one extreme of getting a couple of monitors connected to a simulator and placed in a conference room to a full-fledge replica of the entire (transmission and/or distribution) control room along with all of the console desks, mapboards, phone systems, and so on.

Table 12.1
Illustrative Set of Training Activities for a Transmission System Operator

Phase	Activities
Orientation phase	Orientation to system operation center Introduction to power system operation Overview of the basic principles of operating a power system Overview of all generation stations and the transmission system.
Basic training phase	Overview of the system operation center, including the hardware, software, communications, and the user interface
Power system control principles phase	AC power applications Electrical workers safety practices Power system facilities Power system control principles: Generation ·Generation control ·Energy interchange ·Hydro and pumped storage operation ·Nonutility generation ·Energy resource planning Power system control principles: Transmission ·MW flow controls and limits ·Voltage controls and limits ·Switching and clearance procedures
Normal operations phase	Normal operations: Generation Normal operations: Transmission ·Security analysis and outage planning ·Switching operations ·Transmission system monitoring and control
Off-normal and emergency operations phase	Transmission contingencies Generation contingencies Extreme minimum and peak loads System dynamics Load curtailment and system restoration

All of these cost a lot of money and time to maintain and the benefit of having them depends on whether the utility sees a major issue around system operator training or not.

12.6 Training Simulators as a Real-Time Simulation Platform

Training simulators both at the transmission and the distribution level have a couple of important and common sets of characteristics. They deliver a high-fidelity simulation of the response of the power system to various stimuli and given their time-dependent models, can also go faster than real time. Leveraging these two key characteristics, one could easily visualize the training simulator delivering one a whole host of new capabilities beyond training to support the system operator (the trained one) in their normal day-to-day job. Examples abound in this area where the operator could really use a real-time simulation tool to check things out prior to actually implementing them in the real field.

- *For the transmission system,* a good example could be in the use of a large wind farm coming on board. We all know they come on suddenly and stop as suddenly, causing major problems to the stability of the grid. However, under different circumstances, if the operator runs simulation of the wind farm coming online, they can see the impact in a safe environment, thereby leading them to try different mechanisms to control the outcome before it actually happens.
- *For the distribution system,* a good example would be analyzing the approach of a storm on the system. One could easily overlay the path of the storm on the system, look at the potential areas of damage, and get ready in advance for it either with the predispatching of trouble and other field crews across the jurisdictions or perform advanced switching maneuvers to mitigate the problem.

These two examples highlight the potential for using the training simulator not just for training but as a support tool for the system operator in real time.

12.7 Training Simulators in the Future

The power system is ever-increasing in complexity. The advent of deregulation has made a huge impact to the job of the transmission system operator. Similarly, the smart grid is poised to do the same to the distribution system operator.

The advent of the smart grid is bringing new types of control mechanisms in the field along with different sensing mechanisms as well. With this ever-increasing complexity, system operators need more advanced training in the areas of power flow, voltage and VAR control/dispatch, contingency analysis, and network analysis. While in the past much of these were in the realm of the transmission operator, some are slowly creeping into the distribution scene as well. As systems are updated and these advanced applications are included in new EMSs and DMSs, the operators need to be trained in how to respond to these new stimuli.

In addition, distribution is poised to be dramatically changed over the next few years. Many of the sensors and controls that are presently being piloted will be rolled out all across the system, new application subsystems will be added to the DMS and possibly other systems in the field, and new regulations will come in and modify how the operators can use some of the new technologies in the field. Last but not least, the advent of AMI is well situated to precede the launch of widespread retail markets in several jurisdictions. As all of this happens, training becomes much more important. This is where a training simulator stands out over all other forms of training.

System operators can receive training that can then be applied to the simulated power system. The effects of what they have learned become much more apparent with this type of hands-on activity. This also allows them the means to try new approaches to problem areas of the power system and find the best way to deal with these problems. The training simulator is the best tool for giving the system operator confidence and experience in the use of new tools to better manage the power system.

End Notes

[1] Demand response or demand side management.

[2] Time of use rates for the residential customer based on availability of interval data from smart meters.

[3] Renewable Portfolio Standards (RPS) are regulations that are enacted in various states in the United States with a view to increasing the production of energy from renewable energy sources, such as wind, solar, biomass, and geothermal. Most times, energy produced from hydro sources is not included in this list. Even though RPS is mainly an American term, similar terms exist in the United Kingdom and other parts of the world.

13

Conclusions and What Is Coming Next on the Horizon

The utility industry is embarking on a transition whose end isn't fully understood. Several new areas are just beginning to blossom. Distributed renewables (mainly solar) and new storage technologies are being developed and brought to the market, demand response (even though its acceptance is still a little slower than anticipated), and micr-grids and their proliferation in college campuses and military bases are all just a few examples of these changes, all of which could have significant impacts on how the grid is managed and operated.

This uncertainty is also fueled at least in part due to the fact that many utilities and regulators aren't yet comfortable relying on smart grid and demand response technologies as a substantive resource planning tool. In short, many people don't believe demand management can accomplish more than a very small and incremental dent in the overall load and system usage profile, and that dent will be accomplished at a high cost and perhaps as a notable security risk compared to simply building more power plants and stringing more wires. In addition, the notion of expanding power capacity simply by building new power generators can't be offered as a reasonable alternative, as obtaining building permits for these new facilities is increasingly difficult due to environmental reasons even though new and more vast sources of fuel are being discovered.

Finally, consumers and stakeholders are pressing for productivity increases to accommodate demand growth and rising capital costs. Users are expecting quality, reliability, and power production increases on the one hand, while at the same time demanding that the electric power industry reduce or mitigate its carbon emissions and increase energy efficiency. Users are also adding new types of loads and generation sources on the grid at their homes with the eventual intent to either become net-zero from a grid perspective or significantly reduce energy drawn from the grid. The new types of loads include PHEV/PEVs and new generation sources are most often of the solar/PV type. This means that while they may be still connected to the grid, their main intent is to draw load from the grid sometimes and deliver supply to the grid at other times.

We can summarize the global energy problem with four main points:

1. *Diversification of energy supply and reserves.* While new and additional sources of petroleum-based resources are being found, they are increasingly being found in difficult environments, are costly to extract to a useable form, and are having more significant environmental problems that need to be solved before they reach the market. This fact combined with the increasing global competition for energy resources (from countries like India and China) and increased risk of importing energy from countries who could use this as a leverage for political reasons, requires a rethinking of how the energy needs are met and are thought of in terms of national security. All of this is resulting in a greater focus on new and alternative sources of energy, some conventional and some unconventional/renewable.

   In the near term, two key factors that are affecting the price of electricity at least in the United States are the low cost of natural gas and the dependence of renewable energy (especially wind) on tax credits. It is unknown how long these will last or what impact they will have on the long-term prices of electricity, and will be something for us to watch.
2. *Increasing focus on climate and environmental change.* While the global dialog and rhetoric on climate change has pretty much slowed down to a whisper, it has moved to a different battlefield and that is on the environmental side of the argument. New coal-fired plants are beginning to get difficult from a permitting perspective and most states and a few other countries are mandating that a nontrivial portion of their supply come from renewable sources.
3. *Increased electric power intensity of the economy.* As industrialized societies continue to grow, worldwide electricity demand is estimated to double by the year 2030, and the minor inconveniences that customers currently notice in the power grid will increase, becoming more pronounced and problematic if no action is taken. Much of this new demand is not just coming from developing and underdeveloped countries that are all electrifying at a rapid pace because they have all identified this as a key path to progress. It is also coming from developed countries because of the increased proliferation of multiple TVs, smartphones, computers, and a plethora of new devices in the home.

In addition, the expected move towards the electrification of transportation will only serve to exacerbate the problem, although this may happen faster with commercial vehicles than with residential vehicles.

Utilities are looking at this increased load and looking at intelligent ways to deliver to this requirement. Coupled with the increases in generation from renewable sources—some distributed and some centralized but all of them volatile in their delivery process—this is forcing utilities to come up with imaginative solutions through energy efficiency and capacity management solutions.

4. *Increasing pressure for infrastructure renewal.* All the changes listed above are resulting in a tremendous pressure to rethink the renewal of energy infrastructure in ways that have never been imagined before. The old paradigm of centralized upgrade of the transmission and distribution infrastructure is being completely rethought to meet this completely new mix of generation/supply and consumption/load. In fact, the advent of microgrids is even bringing more attention to the transportation of alternative fuels like natural gas that could be used to generate electricity locally.

   An important and critical aspect of this infrastructure revitalization is the focus on substation and feeder automation. For the first time in the history of electricity, a lot of intelligence is being added to the distribution grid. This is being done through the placement of sensors and controls in the field supported by local intelligence to actually make localized decisions. This will drive a change from today's mostly centralized systems in which all the data comes to a single system (like EMS, DMS, and OMS) to a more distributed system in which much of the analysis and decision may take place in a distributed manner and only some of the effort done centrally.

All of this will make managing and operating the grid much more complex. What does this mean for the system operator of the future?

We will now end this book with a series of predictions on the changes that will come the way of the system operator of the future.

- In transmission systems, more and more PMUs will be installed over time, leading to a much more accurate understanding of the state of the power system, which will in turn lead to a dramatic redesign of the network application suite, in which one may question the long-term role of the state estimator at least in its present form.

  The advent of PMUs will also provide the system operator with predictive tools that will alert the system operator to potential problems that may impact the grid in an adverse manner, provide them with multiple options to solve the problem in an optimal manner, and give them enough time to respond.

The advent of distribution PMUs that are coming preinstalled in the relays and other devices of tomorrow will change the distribution system operator's capabilities in ways that we cannot even imagine today.

- The vast amounts of data coming into the utility systems—some in real time and some slower—will lead to a greater reliance on decision-support systems in which analysis results will be fed to the system operator instead of today's data-based flow of information. This will allow the system operator to make better and more optimal decisions in a more time-effective manner. This aspect applies to both transmission and distribution.
- There will be larger amounts of distributed control all the way from the substation/feeder level to the customer's premise level.

The newer devices like PEV, distributed storage, and others will allow the provision of new and imaginative controls that may be leveraged by the system operator through means like V2G and so on, in which thousands of distributed controls can be brought to bear in running the system in an optimal and cost-effective manner.

This could also come from increased availability and use of HVDC circuits and microgrids, which allow for an increased level of localized control both at the grid level as well as at the energy supply level. This means that the system operator may need to become more nimble and respond to new and different types of stimuli that may impact the grid. The newer tools will need to be more intelligent and automated, leading to the development and deployment of intelligent decision support systems. It can be anticipated that these systems will first show up at the microgrid level and then move up the chain to the full transmission and distribution grid levels.

- Given the increased focus on controls and sensing at the premise level, it will be accompanied by an increased focus on privacy. It is anticipated that (at least in the United States and some of the more developed countries), there will be specific laws on how this information can be collected, what form it can be stored, and how this information can be used. This is new territory for the utility industry and they will slowly learn how to continue to operate seamlessly under this new paradigm.
- Given the increased number of sensors and controls in the grid, leading all the way into the residential premise, it is natural that cybersecurity

will take on levels of importance somewhat similar to how financial data is treated by the brokerage houses of today. The present set of rules governing cybersecurity, which predominantly cover only the transmission system, will extend to beyond the distribution grid all the way to the home. This aspect will change or have a serious impact on just about everything in how the grid is being managed and controlled, although much of this may happen in the background within each vendor's offerings.

- With wholesale markets being the norm in some form in much of the world; it is only natural that retail markets will be the next horizon to cross in terms of the commercial side of utilities and their way of operating. While states like Texas and a few others have already experienced this, a large-scale move towards retail choice will change much of how the system operator manages and operates the system because then even at the distribution level, the utility will have to depend on the availability of the information from other third-party sources, many of whom may not have the rigor and reliability as themselves.

  Even at the wholesale level, the advent of PMUs supported by WAMS applications leads to a requirement for increased cooperation between RTOs. WAMS systems can support very effectively the ability to perform stability and reliability analyses over a wide area encompassing several RTOs and even be somewhat predictive in nature. This capability will become more critical as the operations of the electrical network will get closer to their limits as a result of increasing load supported by fewer large centralized generation plants and more by distributed/renewable sources of supply and more control of end-use load as well. The present discussion of seams issues, which is generally done one-on-one between RTOs will take on greater importance with WAMS systems, as they need to be supported by broader regional cooperation and coordination not just between RTOs but also between utilities in ways that has not yet been considered.

One question that gets asked very often is, "If one were to start with the design of system operations from scratch today, keeping in mind where the future is leading us, how would we do this differently?" The fundamentals of the system of the future will be different. Given some of the key changes that the future is bringing upon us—microgrids, distributed renewables, distribution and transmission automation, PMUs, storage, and so on, the answer can be given as follows:

There is still a need for systems like EMS, DMS, DEMS, and so on. The future of the OMS can be in doubt especially if the DMS can perform much if not all of its functions with the business interfaces back into the rest of the utility's back-office systems. These systems will need to become extremely well integrated with each other and be able to move most important information and control between each system back and forth.

However, in addition, given that the today's distinct line between the functionality in transmission and distribution will become blurred, a need for a system/regional outlook of the grid will become more important. This new system, whether it is being run at a RTO level or even above them at a regional level, will focus on performing regional level stability/security/congestion analysis of the system using PMU (and other) data supported by more sophisticated WAMS-like applications, which then become responsible for keeping the regional backbone of the larger electric grid.

This new system will also be integrated with all the systems described above, across the utilities in the region in addition to being in constant contact with the various distributed and automated controls available to it. It is still unknown as to who or what entity will be running this new system or what level of control they will have to effect the movement of the system in the right direction. However, it is critical that a system of this kind is somewhat necessary given the regional nature of some of the latest disturbances (e.g., the 2003 Northeast blackout). Add to the mix the broader movement of supply away from centralized to more of a distributed model and also the advent of smart grid and its associated technologies and systems.

All of these discussions lead one to looking at system operations from an entirely different lens. However, whatever direction this area moves in, the future of the system operator is very bright. New problems will be identified, new challenges will appear on the horizon, and for all of these new solutions will be developed, some technological in nature and others process-oriented. However, change will be constant and the level of excitement will always be high.

Acronyms and Abbreviations

AC	alternating current
ACE	area control error
AGC	automatic generation control
AMI	automated metering infrastructure
AMR	automated meter reading
ATC	available transmission capacity
CA	contingency analysis
CAIDI	Customer Average Interruption Duration Index
CAIFI	Customer Average Interruption Frequency Index
CAISO	California Independent System Operator
CES	community energy storage
CIP	critical infrastructure protection
CIS	customer information system
CPS1 CPS2	Control Performance Standards 1 and 2
CT	current transformer
CVR	conservation voltage reduction
CVVC	conservation volt-VAR control
DC	direct current

DEMS	distributed energy management system
DER	distributed energy resource
DMS	distribution management system
DR	demand response
DSCADA	distribution SCADA
DTS	dispatcher training simulator
ED	economic dispatch
EMS	energy management system
EPA	Energy Policy Act
ERCOT	Electric Reliability Council of Texas
ERO	Electric Reliability Organization
ERP	enterprise resource planning
ESP	energy service provider
ETOR	estimated time of restoration
FCI	faulted circuit indicator
FEP	front-end processor
FERC	Federal Energy Regulatory Commission
FLISR	fault location identification and service restoration
GHG	greenhouse gas
GIS	Geospatial Information System
GMS	generation management system
GPS	Global Positioning System
HAN	home area network
HTS	high-temperature superconductors
ICCP	Inter Control-Area Communication Protocol
IED	intelligent electrical device
IOU	investor owned utility
IPP	independent power producer
ISO	independent system operator

ISO-NE	independent system operator-New England
ITC	independent transmission company
IVR	interactive voice response
IVVC	integrated volt-VAR control
KCL	Kirchhoff's current law
KPI	key performance indicator
kV	kilovolts
KVL	Kirchhoff's voltage law
kWh	kilowatt hours
LSE	load serving entity
LTC	load tap changer
MAIFI	Momentary Average Interruption Frequency Index
MDM	meter data management system
MISO	Midwest Independent System Operator
MVAR	megavolt-ampere reactive
MVARH	megavolt-ampere reactive hour
MW	megawatt
MWH	megawatt hour
NERC	National Electricity Reliability Council
NYISO	New York Independent System Operator
O&M	operations and maintenance
OASIS	open access same-time information system
OKA	OK on arrivals
OMS	outage management system
OPF	optimal power flow
PEV	plugged-in electric vehicle
PHEV	plugged-in hybrid electric vehicle
PII	personally identifiable information
PJM	Pennsylvania Jersey Maryland ISO

PMU	phasor measurement unit
PT	potential transformer
PUC	public utility commission
PUHCA	Pubic Utilities Holding Company Act
PURPA	Public Utility Regulatory Policy Act
PV	photo-voltaic
REP	retail energy provider
REZ	renewable energy zone
ROR	rate of return
RPS	renewable portfolio standard
RTG	regional transmission group
RTO	regional transmission operator
RTU	remote terminal unit
SAIDI	System Average Interruption Duration Index
SAIFI	System Average Interruption Frequency Index
SCADA	supervisory control and data acquisition
SCED	security constrained economic dispatch
SCUC	security constrained unit commitment
SE	state estimator
SEC	Securities and Exchange Commission
SOA	service-oriented architecture
SPP	Southwest Power Pool
T&D	transmission and distribution
TCMS	trouble-call management system
TO	transmission operator
TOU	time of use
TTC	total transmission capacity
TU	Transmission User
UC	unit commitment

UI	user interface
UPS	uninterrupted power supply
V2G	vehicle to grid
VAR	volt-ampere reactive
VFT	variable frequency transformer
VIU	vertically integrated utility
VVO	volt-VAR optimization
WAMS	wide area monitoring system

About the Author

Dr. Subramanian (Mani) Vadari is the founder and president of Modern Grid Solutions, where he consults with smart grid companies (utilities and vendors) in setting the strategic and technical direction for developing key aspects of the generation/transmission/distribution system of the future. Prior to founding Modern Grid Solutions, Dr. Vadari was a vice-president at Battelle, where he led the development of an industry-leading demand management product. Earlier, Dr. Vadari was also a partner at Accenture, where he was one of the lead partners in their global T&D practice having founded their system operations and smart grid practice. Dr. Vadari also previously served as a lead engineer at ESCA (Areva T&D), focusing on power system and deregulation applications and their delivery. While at ESCA, Dr. Vadari developed ESCA's Transient Stability application and also led the industry's leading DTS product team responsible for development, project support, training, marketing, and sales support.

Dr. Vadari brings over 25 years of experience delivering solutions to the electric utility industry, focusing on T&D grid operations, generation operations, energy markets, and smart grid. His experience spans the regulated and unregulated arena for utility and energy companies. His roles have primarily been business architect, and/or solution delivery for many leading utility companies in North America and around the world. Dr. Vadari is considered a smart grid subject matter expert, offering much sought-after perspectives on the entire value chain of an electric utility from generation to consumption. Dr. Vadari has authored over 40 articles in a variety of areas from dispatcher training simulator (DTS) development to artificial neural networks to electricity utility deregulation and the smart grid. A frequent keynoter at industry events in the United States and abroad, he has served on the boards of GridWise Alliance and Smart Grid Roadshow. Dr. Vadari presently serves on the advisory board of several companies.

Index

Recent Artech House Titles in Power Engineering

The Advanced Smart Grid: Edge Power Driving Sustainability, Andres Carvallo and John Cooper

Battery Management Systems for Large Lithium Ion Battery Packs, Davide Andrea

Designing Control Loops for Linear and Switching Power Supplies: A Tutorial Guide, Christophe Basso

Electric System Operations: Evolving to the Modern Grid, Mani Vadari

Energy Harvesting for Autonomous Systems, Stephen Beeby and Neil White

Power Line Communications in Practice, Xavier Carcelle

For further information on these and other Artech House titles, including previously considered out-of-print books now available through our In-Print-Forever® (IPF®) program, contact:

Artech House
685 Canton Street
Norwood, MA 02062
Phone: 781-769-9750
Fax: 781-769-6334
e-mail: artech@artechhouse.com

Artech House
16 Sussex Street
London SW1V 4RW UK
Phone: +44 (0)20 7596-8750
Fax: +44 (0)20 7630-0166
e-mail: artech-uk@artechhouse.com

Find us on the World Wide Web at: www.artechhouse.com